AURORA
DOVER MODERN MATH ORIGINALS

Dover Publications is pleased to announce the publication of the first volumes in our new Aurora Series of original books in mathematics. In this series we plan to make available exciting new and original works in the same kind of well-produced and affordable editions for which Dover has always been known.

Aurora titles currently in the process of publication are:

Optimization in Function Spaces by Amol Sasane. (978-0-486-78945-3)

The Theory and Practice of Conformal Geometry by Steven G. Krantz. (978-0-486-79344-3)

Numbers: Histories, Mysteries, Theories by Albrecht Beutelspacher. (978-0-486-80348-7)

Elementary Point-Set Topology: A Transition to Advanced Mathematics by André L. Yandl and Adam Bowers. (978-0-486-80349-4)

Additional volumes will be announced periodically.

The Dover Aurora Advisory Board:

John B. Little
College of the Holy Cross
Worcester, Massachusetts

Ami Radunskaya
Pomona College
Claremont, California

Daniel S. Silver
University of South Alabama
Mobile, Alabama

Numbers

Histories, Mysteries, Theories

Albrecht Beutelspacher
Justus Liebig University Giessen

TRANSLATED BY
Andrea Bruder, Andrea Easterday, & John J. Watkins

DOVER PUBLICATIONS, INC.
Mineola, New York

Bibliographical Note

This Dover edition, first published in 2015, is a new English translation of
Zahlen: Geschichte, Gesetze, Geheimnisse, first published in Germany by C.H.
Beck, in 2013. It is published by special arrangement with Verlag C.H. Beck
oHG, Wilhelmstr. 9, 80801 Munich, Germany. For the English edition, the
translators and the author have added a section titled "Additional Notes," which
begins on page 89.

Library of Congress Cataloging-in-Publication Data

Names: Beutelspacher, Albrecht, author.
Title: Numbers : histories, mysteries, theories / Albrecht Beutelspacher ;
    translated by Andrea Bruder, Andrea Easterday, John J. Watkins.
Other titles: Zahlen. English
Description: Dover edition. | Mineola, New York : Dover Publications, 2015. |
    Series: Aurora: Dover modern math originals | Originally published in
    German: Zahlen : Geschichte, Gesetze, Geheimnisse (Munich : C.H. Beck,
    2013). | Includes bibliographical references and index.
Identifiers: LCCN 2015042269 | ISBN 9780486803487 | ISBN 0486803481
Subjects: LCSH: Numeration—History. | Mathematics—History. | Counting.
Classification: LCC QA141 .B4813 2015 | DDC 513.5—dc23 LC record available
at http://lccn.loc.gov/2015042269

Manufactured in the United States by RR Donnelley
80348102   2016
www.doverpublications.com

Contents

Preface

"What exactly is a number?" Hardly anything will embarrass a mathematician more than this simple question. One might think mathematicians should know what a number is. After all, they are concerned with numbers all the time!

Nevertheless, every mathematician pauses at this question and mumbles something along the lines of "This is not as easy to answer as you might think." After a while, they have to admit that they do not really have an answer to this question. They may even be tempted to refuse to answer the question altogether.

Scandalous. The simplest of mathematical questions remains unanswered!

The reason that the question remains unanswered is clearly that there is no answer. In any case, not a simple one. And not just one. This little book is not only an attempt to explain what a number is but also to posit an answer as to why there *is* no simple answer to the question "What is a number?"

Throughout the history of mathematics, of course, there have been many attempts to describe what a number is.

- The mathematicians of ancient Greece said: Numbers are the natural numbers $1, 2, 3, 4, 5, \ldots$. The Greeks at least knew that there are numbers without end.

- Soon they realized that fractions of numbers were needed. One had to split objects in halves or into three equal pieces; this is how the necessity for numbers such as $1/2$ or $2/3$ emerged. Because they used them in their daily business, the merchants of medieval times were convinced that fractions were numbers, too.

- It took much longer to accept negative numbers such as -1 or -5, and even 0, as numbers.

- Greek mathematicians had already stumbled across numbers such as $\sqrt{2}$, $\sqrt{7}$, and $\sqrt[3]{2}$. Some of these numbers "exist" because they may be interpreted geometrically as the length of a line segment. However, as numbers per se, roots initially presented a challenge.

- Then some mysterious objects appeared, such as the circle number π or Euler's number e. These numbers are reminiscent of long dormant meteors found on Earth. One can examine and try to understand them, but senses that these numbers are messengers announcing distant worlds, representatives of the vastness of the mysterious real numbers.

- The "imaginary unit" i, which is defined as the square root of -1, finally marked the end of the fun, even for many mathematicians.

It makes us dizzy. So many different kinds of numbers! We scratch our heads and wonder:

- Will this never end?

- Do we need all of these numbers?

- Does the definition of what a number is depend on when that definition was made?

Apart from some answers to the initial question of what a number is, this book also shows:

- the wealth of experiences that numbers have to offer,

- that many things can be described by numbers,

- which numbers are particularly fascinating, and

- how numbers still hold many mysteries today.

Chapter 1

Natural numbers

> *Numbers are free creations of the human mind,*
> *they serve as a means of apprehending more easily*
> *and more sharply the dissimilarity of things.*
>
> Richard Dedekind
> *Was sind und was sollen die Zahlen?* (1888)

1.1 Counting

We humans are born with a number sense. However, this innate ability is not unique to humans, since animals also have a sense of numbers. Many higher developed animals have been shown to be able to handle numbers: They are capable of recognizing equal numbers of things and they can distinguish between larger and smaller quantities. Apparently this is an advantage in the fight for survival.

Animals have a sense of numbers, but they are unable to count. This activity is coupled with language and thus reserved for humans. However, we are not born with this ability. Every human must learn how to count. A newborn's understanding of numbers does not significantly differ from that of a chicken. We arrive in this world with the ability to distinguish certain amounts by their size. Everything else we must make an effort to learn.

If we do not learn, we remain at "one, two, many." Again and again there have been reports of "primitive" cultures that only have words for the numbers one and two.

1

We can determine small numbers at a glance. We can tell how many objects there are without counting if there are five or fewer objects. If there are more objects, this is only possible if they are arranged in a pattern.

With the development of the first human settlements, it became important to accurately record larger numbers. There is some evidence of numerical representations as old as 20,000 to 30,000 years: Bones have been found with scratch marks that represent numbers. The first number systems allowing large numbers to be recorded originate from the Babylonians (c. 2000 B.C.; see Chapter 2).

Probably even older than the representation of numbers is the activity of counting. Counting is based on the perception of the world's rhythms and how they are captured through language. Life is structured by many uniformly repeating processes. Among these are the continuous change of day and night, the rhythm of the seasons, the sequence of steps while walking, and the beat of the heart.

I imagine that humans at some point began to notice the rhythm of walking by singing, speaking, or drumming along. Perhaps they initially just thought aloud their words for left and right or one and two, perhaps they sang along, perhaps someone started drumming a two part beat—no matter how they did it, it was the beginning of counting. Or at least the beginning of counting to 2: one, two, one, two, and so on.

At some point, someone had the crazy idea to break this closed cycle and to simply keep counting: one, two—three. Probably a beat of four was next: one, two, three, four, one, two, three, four, and so on.

And when the first hurdle is taken, it is easy to conquer the second one as well. With three, an infinite sequence of numbers was born: one, two, three, and so on. Then it was clear that one could always take another step and that beyond every number another number follows. This was the real beginning of counting. The one who can say three can count.

The mathematical image for this infinite sequence is the number line, more precisely, the number ray. This is an infinitely long ray, which begins at 0 and along which we move step by step to conquer all of the numbers: Beginning with the first step, we arrive at one from zero, the second step takes us to two, the next step to three, and so on. The numbers reached in this manner—the numbers $0, 1, 2, \ldots$— are called the *natural numbers*.

The number ray

The number line, that is the number ray extended by the negative numbers, includes the *whole numbers*, also called *integers*. The number line is a means for explaining and illustrating many properties of numbers.

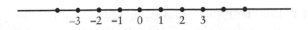

The number line

One may move along the number line in small or in large steps, one may jump forward and backward, or one may try to close the gaps between the whole numbers. We will return to the idea of the number line throughout this book.

1.2 Properties of numbers

At around 600 B.C. a decisive event happened in European intellectual history, with consequences that cannot be overestimated: People in Greece and Asia Minor discovered the power of thought. More precisely, they discovered the possibility of abstraction. This means the goal of understanding is reached through thought and simplification. The advantage of this sometimes tedious process is that one can examine abstract objects and circumstances through thought, and in particular, through logic. If one no longer ponders a ditch in the sand, but instead an abstract line, if one no longer sees seven dogs dashing about in play, but instead the number 7, then one can logically relate these objects to other objects and draw conclusions about them which have a degree of certainty that can never be obtained in the empirical world.

Two people who crucially shaped this "first Enlightening" are Thales of Miletus (c. 624–546 B.C.) and Pythagoras (c. 570–510 B.C.). Pythagoras in particular had a great impact on the development of Greek mathematics through his school in Croton in southern Italy.

For the Pythagoreans, numbers were not only a means of quantifying things, they also distinguished between certain "qualities" of numbers. For example, they defined even numbers: A number is even when it is divisible by 2 without leaving a remainder. This is probably one of the earliest mathematical concepts in history to be defined.

The Pythagoreans also knew and phrased laws such as *odd plus odd is even*. This means: The sum of any two odd numbers is always an even number. For example, $3 + 5$ equals 8, an even number. What is crucial here is that this is not only true for the numbers 3 and 5, but for *all* odd numbers. Mathematicians call such a statement a "theorem."

Number symbolism and number mysticism were central elements of Pythagorean number theory. The Pythagoreans assigned certain non-mathematical properties to certain kinds of numbers and obtained an ordering principle which they used to interpret the world. For example, odd numbers were "masculine," even numbers were "feminine." At the same time, even numbers represented the unbounded, while odd numbers represented the bounded and limited.

A crucial tool for finding properties of numbers, representing them, and establishing relations between their properties was the *figurate numbers*. The idea is to arrange a number of pebbles into a geometric figure and to study the number of pebbles used. For example, one can represent an even number as a rectangle, where one side consists of two pebbles.

An even number

An odd number is represented by using an additional pebble.

An odd number

From this representation one sees immediately that an odd number is obtained if one adds 1 to or subtracts 1 from an even number. The rule of "odd plus odd is even" is now immediately clear: One "slides together" two odd numbers, and the result is automatically an even number:

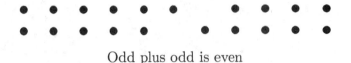

Odd plus odd is even

The *square numbers* were of particular significance. These are numbers that can be arranged in a square.

The first few square numbers

The first few square numbers are $1, 4, 9, 16$. Their representation as a figurate number also makes clear how one obtains the next square number:

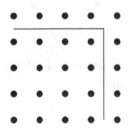

Square numbers and odd numbers

We see that we can add an odd number to a square number in order to obtain the next square number. For example, $4^2 + 9 = 16 + 9 = 25 = 5^2$. Indeed, the differences of consecutive square numbers are $3, 5, 7, 9, \ldots$. In general: We obtain the nth square number by adding the nth odd number to the previous square number. In other words: The sum of the first few odd numbers is a square number; for example, $1 + 3 + 5 + 7 + 9 = 25$.

Just as interesting as the square numbers, and in some sense even simpler, are the *triangular numbers*, the numbers needed in order to form a triangle.

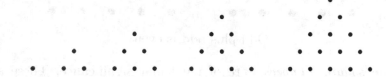

The first few triangular numbers

The sequence of triangular numbers begins with $1, 3, 6, 10, 15, \ldots$. The nth triangular number is obtained by increasing the previous triangular number by n. In other words: The sum of the first n natural numbers is a triangular number. For example, $1 + 2 + 3 + 4 + 5 = 15$.

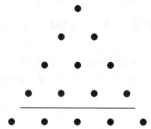

Consecutive triangular numbers

Of the utmost importance to the Pythogoreans was the *tetrad*, the fourth triangular number.

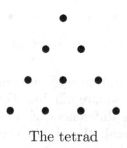

The tetrad

Here, three important concepts come together. The triangular shape, the sum of the first four numbers $1 + 2 + 3 + 4$, and the result, the number 10. The number 10 was considered perfect because "it includes the entire nature of numbers."

We might dismiss this way of obtaining insight as a childish pebble game, however, failing to recognize the significance of this method. In reality, these "proofs without words" are the first attempts at discovering patterns in the realm of numbers. This must especially be appreciated because there was no formalism available to the Greeks with which they could have described the patterns they studied.

1.3 Magic squares

Magic squares are at the threshold between a pre-scientific "magical" view of numbers or symbols and their rational perception. One may view a magic square as a combination of number symbols that yield a coherent overall picture, as well as a number scheme whose structure is determined by arithmetic operations, for example addition.

The oldest magic square is the Chinese Luoshu, which supposedly was found more than 4000 years ago carved into the shell of a tortoise.

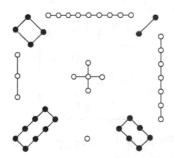

The Luoshu magic square

Mathematically, it only depends on the numbers in each smaller square; in mathematical abstraction, the Luoshu magic square looks like this:

4	9	2
3	5	7
8	1	6

A 3 × 3 magic square

The numbers in the square are added in each row, column, and along the diagonals. The "magic" is that we obtain the same sum every time! In the Luoshu magic square, the "magic sum" is always 15. For example, we compute the following sums along the rows: $4+9+2 = 15$, $3+5+7 = 15$, $8+1+6 = 15$.

Constructing a *magic square* consists of arranging the numbers 1 to 9, 1 to 16, or, more generally, 1 to n^2, into a square pattern in such a way that the sum of the numbers in each row, column, and diagonal is the same.

It is much harder to construct magic squares with many rows and columns. But mathematicians have proved that for each natural number n that is at least 3, there exists an $n \times n$ magic square. Moreover, as n increases, there are more and more magic squares.

A magic square that appears in Albrecht Dürer's mysterious etching *Melencolia I* in 1514 has attracted particular attention. Above an angel's head one finds a magic square made up of the numbers 1 to 16. The magic sum is 34.

16	3	2	13
5	10	11	8
9	6	7	12
4	15	14	1

Dürer's 4×4 magic square

Not only does the magic sum appear in the rows, columns, and diagonals, it also occurs in the 2×2 squares in the corners, in the numbers in the four corners, and in the four numbers in the middle.

Another subtlety is that on the bottom in the middle there are the numbers 15 and 14, the year the etching was made, and that in the lower corners the numbers 4 and 1 are found, which correspond to the letters D and A, Albrecht Dürer's initials.

1.4   Prime numbers

In 1960 the Belgian archaeologist Jean de Heinzelin de Braucourt found a bone in the remains of a fishing village in Ishango, now on the border between contemporary Uganda and the Democratic Republic of the Congo. The bone Heinzelin found is about 10 centimeters

Albrecht Dürer, *Melencolia I*

long, and is now believed to be 20,000 years old (see Section 2.1). Three rows of numbers were notched into the bone by the stone age inhabitants of Ishango. One row of notches fascinates modern scholars because it displays the numbers 11, 13, 17, and 19. It is difficult not to think of prime numbers when considering the Ishango bone, because 11, 13, 17, and 19 are the primes between 10 and 20. Whether a stone age human was able to understand and use prime numbers remains a mystery.

Greek mathematicians were familiar with prime numbers by about 500 B.C. In order to understand the natural numbers by envisioning them as figurate numbers, one must consider rectangular numbers. Rectangular numbers can be displayed in the form of a rectangle with edges at least 2 units in length. For example, 15 is a rectangular number because it can be displayed as a 3×5 rectangle. Prime numbers are those numbers that cannot be represented by a rectangle. In other words, prime numbers are numbers that can only be displayed as a line. For example, 7 is a prime number, because one cannot display seven objects in a rectangle with one side of length 2, 3, 4, 5, or 6.

Prime numbers were defined in the first (and most important) book in mathematics, Euclid's *Elements* (c. 300 B.C.). In Euclid's formulation the notion lingers that "a prime number is a number that is not a rectangular number." He further explains, "a prime number is a number that can be measured only by unity."

Prime numbers are easily defined, and examples are easy to find, but they are nonetheless among the most fascinating and mysterious concepts in mathematics. The history of the study of prime numbers is one of many mathematical stories that began in ancient Greece and is not yet finished today.

The first prime numbers are 2, 3, 5, 7, 11, 13, 17, 19, 23, 29, 31. There are also large primes such as 65,537. The largest prime number known today is $2^{57885161} - 1$, which has over 17 million digits.

Today one defines a *prime number* as a natural number larger than 1 that has only 1 and itself as divisors. The emphasis here is on "only," because every natural number has 1 and itself as divisors but prime numbers have *only* these two divisors. Prime numbers thus have as few divisors as possible.

Despite this negative definition, prime numbers have tremendous significance. In some sense prime numbers are the most important natural numbers, because all other numbers can be represented using primes. In fact, the Fundamental Theorem of Arithmetic guarantees that *every natural number larger than 1 is either itself a prime number or else is a product of primes.* For example, one can convince oneself the following factorizations are correct:

$$
\begin{aligned}
24   &= 2 \cdot 2 \cdot 2 \cdot 3 \\
999 &= 3 \cdot 3 \cdot 3 \cdot 37 \\
1000 &= 2 \cdot 2 \cdot 2 \cdot 5 \cdot 5 \cdot 5 \\
1001 &= 7 \cdot 11 \cdot 13
\end{aligned}
$$

The primes are the atoms of the world of numbers. Just as every chemical compound is assembled from precise quantities of specific elements, every natural number is assembled from precise quantities of specific prime numbers. For example, every water molecule is assembled from two hydrogen atoms and one oxygen atom and is therefore denoted H_2O. Analogously, the number 999 is assembled from three 3s and one 37, while in 1001 the prime numbers 7, 11, and 13 each appear exactly once.

In contrast to chemistry, in mathematics there are infinitely many "atoms": the prime numbers. Euclid states the infinitude of primes as Theorem 20 in Book IX of the *Elements*. This theorem demonstrates Euclid's mastery with a visionary assertion, a brilliant proof, and judicious manipulation of the concept of infinity.

Euclid's careful formulation states: *There are more primes than any assigned quantity of primes.* That means there is no largest prime number, because up to a hypothetical largest prime number there is an assigned quantity of primes, and Euclid's theorem states that there are more primes than any assigned quantity. Today we simply say "there are infinitely many primes."

Euclid's proof is based on a brilliant idea. Imagine a finite quantity of prime numbers $p_1, p_2, \ldots, p_n$. We must show that there exists at least one additional prime number. Now comes the brilliant idea: Euclid considered the number

$$m = p_1 \cdot p_2 \cdot \ldots \cdot p_n + 1,$$

which is the product of this list of prime numbers, plus 1. The "plus 1" is the real show stopper. Either the number m can be divided by one of the prime numbers $p_1, p_2, \ldots, p_n$ or else m is itself a prime number, in which case m is a new prime. If m is not prime, it has a prime divisor p. If p were for example equal to p_1, then p would be a divisor of the number m, and also a divisor of the number $p_1 \cdot p_2 \cdot \ldots \cdot p_n$, because $p = p_1$. Thus p would be a divisor of the difference between the numbers m and $p_1 \cdot p_2 \cdot \ldots \cdot p_n$, which is the number 1. This contradiction demonstrates that p cannot be any of the prime numbers $p_1, p_2, \ldots, p_n$, and p is therefore the *additional* prime number we seek.

Although there are infinitely many prime numbers, one has the impression that with larger numbers it becomes ever more difficult to find which ones are prime. This impression originates in part because the primes are distributed in a highly irregular manner. On the other hand, there are surprisingly many primes. To this end, the great German mathematician Carl Friedrich Gauss (1777–1855) established a precise conjecture. Gauss conjectured the number of primes less than or equal to a certain number n can be estimated using the expression "n divided by the number of digits of n." Let us consider a few examples:

n	Digits k in n	Number of primes $\leq n$	n/k	$\frac{1}{2} \cdot n/k$
9	1	4	9	4.5
99	2	25	49.5	24.75
999	3	168	333	166.5
9999	4	1229	2499.75	1249.875
99999	5	9592	19999.8	9999.9

One observes astounding agreement between the numbers in the third and fifth columns. In fact, Gauss deduced from such data that the quantity "1/2 times n divided by the number of digits of n" constitutes a good approximation for the number of primes less than or equal to n. More precisely, he conjectured that n divided by the natural logarithm of n asymptotically approaches the number of primes less than or equal to n.

Gauss could not prove this *Prime Number Theorem*. The first mathematicians to successfully prove the theorem were the Frenchman Jacques Hadamard (1865–1963) and his Belgian colleague Charles de la Vallée Poussin (1866–1962) independently in 1896 and 1897.

The Prime Number Theorem extends far beyond Euclid's insight. The theorem states not only that the prime numbers never cease but also that there are unimaginably many primes. The density of primes lessens, but it lessens very slowly. For example, one might ask how many primes exist that are less than 10^{100}. The number 10^{100} is much larger than the number of elementary particles in the universe, and no one will ever list all the primes less than 10^{100}. However, the Prime Number Theorem tells us how many prime numbers lie so far afield. Quite simply, the number 10^{100} divided by the number of its digits, which is 100, is 10^{98}. About half of these are primes, $5 \cdot 10^{97}$. In other words, since no even number larger than 2 is prime, for such unimaginably large numbers about 1 percent of all odd numbers are primes.

1.5 From Pythagoras to Fermat

Today Pythagoras is both famous and infamous for the theorem that carries his name, the "Pythagorean Theorem." Whether the theorem is rightly named is questionable, for the Babylonians knew of the theorem more than 1000 years before Pythagoras lived and taught. It is, however, likely that Pythagoras was the first mathematician to prove the theorem, or at least that the theorem was proved for the first time in his school. From the time of Pythagoras onwards, the Pythagorean Theorem has been an imperishable mathematical theorem rather than a law of nature.

Pythagoras found perhaps the first—but certainly not last—proof of the Pythagorean Theorem. Today more than 400 proofs are known; among others, Albert Einstein, the German philosopher Arthur Schopenhauer, and the former American president James A. Garfield have proved the theorem.

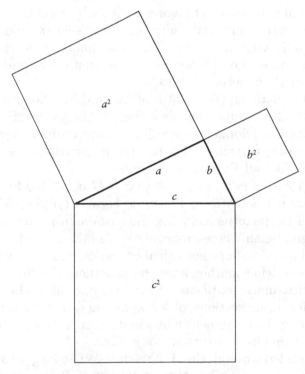

The Pythagorean Theorem

The Pythagorean Theorem is thus: Let us name the length of the two short sides of a right-angled triangle a and b and the length of the long side c. Then $a^2 + b^2 = c^2$. In words, the sum of the areas of the squares of the short sides of the triangle is equal to the area of the square of the long side.

The Pythagorean Theorem has many applications, in particular in the computation of lengths, but it has also provided meaningful stimuli in number theory. Once again this story begins in ancient Greece and continues to engage us today.

The opening question is very simple: Do right angled triangles exist, the length of whose sides are whole numbers? In the language of numbers, the question becomes: Are there whole numbers a, b, and c, for which $a^2 + b^2 = c^2$? If so, for which numbers is the theorem true?

By trial and error one quickly identifies the numbers $a = 3$, $b = 4$, and $c = 5$. Indeed, $3^2 + 4^2 = 9 + 16 = 25 = 5^2$. Natural numbers a, b, and c that satisfy the equation $a^2 + b^2 = c^2$ are called a *Pythagorean triple*. The numbers 5, 12, and 13 are another Pythagorean triple, because $5^2 + 12^2 = 25 + 144 = 169 = 13^2$. Babylonian mathematicians were familiar with Pythagorean triples by 1500 B.C., and even knew triples that include large numbers such as 56, 90, 106 or 12,709, 13,500, 18,541. One does not find such numbers by chance; the Babylonians must have developed a systematic procedure for the generation of Pythagorean triples.

The mathematician Diophantus of Alexandria, who probably lived around A.D. 250, completely described Pythagorean triples in his book *Arithmetica*. Diophantus specified a procedure by which one can generate not only certain triples, not just many triples, but infinitely many and in fact all Pythagorean triples.

Almost 1500 years later, in the year 1637 or so, the French jurist and amateur mathematician Pierre de Fermat (1607–1665) studied the works of Diophantus including the *Arithmetica*. The term "amateur mathematician" deserves some explanation. While true that Fermat had no formal mathematical education and also never earned his bread practicing mathematics, he nonetheless exchanged letters with the great mathematicians of his time and often challenged his peers by offering suggestions, observations, and speculations. Fermat belongs among the prominent mathematicians of history. To this day he is famous thanks to a certain conjecture.

Fermat waded through the *Arithmetica*. When he got to the point where Diophantus describes the generation of Pythagorean triples, a brilliant idea struck him like a bolt of lightning. Fermat wrote in the margin of the book, "it is impossible to dismantle a cube into two cubes, or a biquadrate into two biquadrates, or in general any power higher than the second into two powers of the same degree. I have found a truly wonderful proof of this, but the margin is too small to contain it."

Using language that today sounds archaic, Fermat asks whether the equation $c^3 = a^3 + b^3$ can be solved by whole numbers (breaking down a cube into two cubes), whether the equation $c^4 = a^4 + b^4$ can be solved by whole numbers (breaking down a "biquadrate," or a power of four, into two biquadrates), or in general, whether the equation $c^n = a^n + b^n$ can be solved for any number $n > 2$ using natural numbers $a, b, c > 0$.

Fermat claimed this is impossible, and the equation has no solution. Fermat believed to have found "a truly wonderful proof," which, alas, was so long he could not write it in the margin of his book.

For centuries the note in the margin gave mathematicians nightmares. At some point, mostly in his or her youth, each and every mathematician tried to find the "truly wonderful proof" of Fermat's Conjecture. Every mathematician tried to prove that the equation $a^n + b^n = c^n$ has no solution involving positive whole numbers for $n > 2$. Everyone tried, without success.

With considerable effort, and not necessarily with "wonderful proofs," special cases were addressed: The equation was proved to have no solution for $n = 3$, for $n = 4$, or for $n = 5$. Some special cases are difficult, others easy. The case $n = 4$ is relatively easy, and even Fermat handled this case explicitly.

Generations of mathematicians attacked the problem, with at most partial success. By 1950 mathematicians knew that no solution exists for n less than 2,000. This means it is at the very least fiendishly difficult to find a solution.

Eventually most mathematicians wrote the matter off. Fermat's Conjecture seemed an unsolvable problem. Those individuals who occupied themselves with Fermat's Conjecture were dismissed as dreamers unable to accept that the problem was simply too difficult.

This was the mood in the mathematical community—until on the 23rd of June, 1993, the bomb dropped. British mathematician Andrew Wiles held a series of lectures at Cambridge University, and at the end of his lectures Wiles claimed he had proved Fermat's

Conjecture. *It is impossible to break down any power higher than the second into two powers of the same degree.* It was exactly as Fermat surmised, but now the conjecture was proven.

Wiles had been fascinated by Fermat's Conjecture for his entire life. However, in contrast to many mathematicians he chose a viable method. He was also willing to engage with the problem for many long years. Even as a newly minted professor Wiles withdrew to work secretly, dedicating all his faculties to the proof of Fermat's Conjecture. He struggled for seven years before achieving success.

In the first manuscript a gap in the proof was discovered, which Wiles was able to address with a new approach.

Wiles' proof is not simple. It is everything but simple. It is long, complicated, and employs the most advanced methods in mathematics. On the other hand, specialists can follow Wiles' proof, and it has been verified many times. The proof is correct. After 350 years, Fermat's Conjecture has finally become a theorem: "Fermat's Last Theorem."

Wiles' proof is certainly not the "truly wonderful proof" that Fermat envisioned (and there is clearly not enough room in the margin to present the proof). Does the "truly wonderful proof" exist? Most mathematicians doubt it, and believe that Fermat was simply mistaken.

Yet the dream remains.

1.6　What are natural numbers?

Humans have worked with natural numbers for millenia. We use numbers to specify quantities and to perform calculations; we investigate their characteristics. However, it never occurred to anyone to ask whether the natural numbers truly exist, or what "existence" actually means. Is there truly an infinitude of natural numbers?

Apparently everyone assumed the natural numbers are $1, 2, 3, \ldots$ (or $0, 1, 2, 3, \ldots$), and put their faith in the three little dots which indicate that the series continues indefinitely.

Does the series continue indefinitely? Good arguments support the claim that the natural numbers do not grow without bound. The largest physically existing number is the number of atoms in the observable universe, or about 10^{80}. One could argue that no larger natural numbers are needed, because there is nothing left to count.

Towards the end of the 19th century it became apparent that these basic principles of mathematics must be clarified. In 1888 the German mathematician Richard Dedekind (1831–1916) published a programmatic script with the title *Was sind und was sollen die Zahlen?* (What are numbers and what should they be?). In the script Dedekind takes a very clear position: "Numbers are free creations of the human mind, they serve as a means of apprehending more easily and more sharply the dissimilarity of things. Through the purely logical structure of number theory and the continuous realm of numbers gained thereby, we can now precisely investigate our conceptions of space and time within this realm of numbers created by the human mind."

In his book Dedekind published a set of axioms for the natural numbers. This set of axioms was adapted by the Italian mathematician Giuseppe Peano (1858–1932) and published in 1889. Today the set is known as the *Peano axioms*. In a certain sense these five axioms "merely" precisely and formally define incremental counting. The first axiom states,

1. *0 is a natural number.*
   The first axiom says there is at least one natural number, because the number zero is a natural number. The next axiom formally states what it means to count:

2. *Every natural number is followed by a successive natural number.*
   This describes a counting increment: With a single step one moves from a number n to its successor, which one denotes with n'.

3. *0 is not the successor of any natural number.*
   This axiom states that zero is not just any natural number, but is instead the first natural number. Like every natural number, zero has a successor, but zero is the only natural number that has no predecessor.

4. *Natural numbers with the same successor are the same.*
   One could formulate the fourth axiom in the following way: Zero has no predecessor, and all other natural numbers have exactly one predecessor. The fifth axiom is the decider:

5. *If a set X of natural numbers contains the number zero and together with each natural number also contains that number's successor, then X is identical to the set of all natural numbers.*

This means that if one starts at 0 and then continues stepwise, one will eventually enumerate every natural number.

One might be tempted to believe that the Peano axioms can be used to construct the natural numbers. One starts with the symbol 0, steps to its successor $0'$, continues to its successor $0''$, thence to $0'''$, and so on. It sounds easy, but there is a drawback: In order to obtain infinitely many numbers in this way one requires an infinite number of apostrophes. The observable universe is not large enough to encompass enough apostrophes, for one cannot draw more than 10^{80} of them. Naturally one might eschew the apostrophes and instead of $0'$ write 1, instead of $0''$ write 2, and so on. Yet then one would need infinitely many symbols $0, 1, 2, \ldots$ and accordingly would require infinitely many spaces in which to write infinitely many numbers.

Spin it as you will, one cannot create infinitely many natural numbers from nothing. The Peano axioms describe the natural numbers, but the axioms do not construct the numbers, and neither do they guarantee the existence of numbers. The mathematician Leopold Kronecker (1823–1891) expressed this in a famous sentence: "God made the natural numbers, all the rest is the work of man."

Let us begin by considering the first part of the sentence. By claiming that God created the natural numbers, Kronecker says these numbers are not the work of man. Mankind cannot fashion the natural numbers. Even in Kronecker's time "God" was a metaphor which signifies that one cannot summon an infinite set from nothing; instead, the set in question must be provided in some other manner. Today one expresses this via an additional axiom that guarantees the existence of the set of natural numbers. This "axiom of infinity" is the admission that things just won't work otherwise. This "axiom of infinity" is the understanding that it is possible to imagine a world in which infinity does not exist: If we want infinity, we must postulate it.

In contrast, the second part of Kronecker's sentence is anything but defensive: "Everything else is the work of man." As soon as we have the natural numbers, the rest goes swimmingly: From the natural numbers one can construct negative numbers, fractions, decimals, roots, logarithms, and generally all other types of numbers.

The first answer to the central question of this book "What is a number?" is the following: If one develops numbers from the experience of counting, that is to say numbers are the abstract model for counting, then "number" means either "natural number" or "whole number." The natural numbers are those that can be counted only in one direction, while the whole numbers model unrestrained counting in both directions.

1.7 Applications: Cryptography

The encryption of secret messages is an ancient procedure whereby the author of a message seeks to achieve confidentiality. The goal is to allow two people to communicate with one another in such a way that no third person can unlock the underlying message using the transferred characters. For this reason the message is encrypted: Sender A encrypts the message and sends the encrypted text to recipient B, who decrypts the received confidential text.

Historically, cryptography has always been performed with the help of a secret key which the sender and the recipient know, but which remains unknown to everyone else. The sender uses the secret key to encrypt the message, and the recipient uses the key to decrypt the message. With the shared secret key, the sender and the recipient defend the privacy of their message from the rest of the world.

Until 1976 cryptographers were convinced that encryption could only be performed using a secret key. The disadvantages of a secret key are obvious—before one can encrypt a message, the secret key must be exchanged. The exchange of keys was achieved primarily through envoys or other similarly insecure methods and was the Achilles' heel of every cryptographic system. In 1976 two young Americans posed a question, which from today's perspective sounds wholly natural, but at the time was downright provocative. Under the title *New directions in cryptography*, Whitfield Diffie (1944–) and Martin Hellman (1945–) asked whether it is possible to communicate confidentially without prior exchange of keys. They imagined the process to be as simple as making a telephone call: One looks up a number filed under the name of the recipient, which is their "public key." With the help of this public key one sends the recipient an encrypted message, which only the recipient can decrypt using their "secret key." Such a system is called a *public key encryption system.*

Diffie and Hellman first made their bold speculation in 1976, but cryptographers could have hit on it earlier, for every collection of mailboxes represents a public key encryption system.

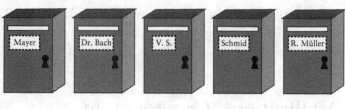

Mailboxes

In particular, every mailbox has a name plate to identify the recipient, a slot for the insertion of mail, and a lock for which only the recipient who owns the mailbox possesses a key. Anyone can send a confidential message to Frau Schmid: One writes the message on a piece of paper and perhaps folds the paper into an envelope. Then one searches for the name of the recipient and tosses the envelope into the corresponding mailbox. The mailbox is equivalent to the public key, and tossing the envelope into the mailbox represents application of the public key. In order to 'decrypt' the message, or in other words, in order to make it readable, Frau Schmid opens the mailbox using her private key, removes the envelope, and can then read the message.

In 1976 Diffie and Hellman had a bright idea, but no solution. They didn't know whether their dream would ever become real. Two years later, Ron Rivest (1947–), Adi Shamir (1952–), and Len Adleman (1945–) realized the dream. Indeed, they discovered that the potential for a public key encryption system was hidden within a famous theorem of number theory. Presumably Rivest, Shamir, and Adleman had no conception of the tremendous potential of their "RSA algorithm." Today the algorithm is widely used and is, for example, employed in every internet encryption.

The theorem on which the RSA algorithm is based is one of several thousand theorems discovered by the Swiss mathematician Leonhard Euler (1707–1783). A simple special case of the theorem is as follows: Consider some natural number m. Raise m to the fifth power, which entails calculating $m^5 = m \cdot m \cdot m \cdot m \cdot m$. Euler's theorem then states that m^5 has the same units digit as m.

Consider this theorem in reverse, and it becomes a magic trick: What is the fifth root of 16,807? If 16,807 is the fifth power of a natural number, then Euler's theorem states that this natural number and 16,807 have the same units digit. The units digit of the fifth root of 16,807 must therefore be 7. A rough estimate shows that the fifth root of 16,807 must be smaller than 10. The mystery number is thus 7. This trick also works if one raises a number to the power of $9, 13, 17, \ldots$, i.e., a power of the form $4k + 1$.

In general, Euler's theorem is as follows: Let n be a product of two prime numbers p and q so that $n = p \cdot q$. (Previously, the number 10 played the role of n; p was 2 and q was 5.) The magic numbers with which we exponentiate m are now $(p-1)(q-1)+1$, or in general some multiple of $(p-1)(q-1)$ plus one, $k \cdot (p-1)(q-1)+1$. (In the above example $p = 2$ and $q = 5$, therefore $(p-1)(q-1)+1 = 1 \cdot 4 + 1 = 5$.) We now choose a number m smaller than n and calculate the number $m^{k \cdot (p-1)(q-1)+1}$. In the above example we were particularly interested in the units digit of this number. The units digit is just the remainder left over after one divides by n. Euler's theorem then states that this remainder is equal to m. In the concise language of mathematics, for every natural number $m < n$ it is true that:

The remainder of $m^{k \cdot (p-1)(q-1)+1}$ after division by n is m.

I suspect that Ron Rivest saw Euler's theorem and thought, m doesn't have to represent a number—it could mean "message." The theorem says one does something rather complicated with m, but in the end m emerges unchanged. Then Rivest had an idea—if one could deconstruct this complicated procedure into two actions and call the first action "encryption" and the second action "decryption" then one has constructed an encryption system. If one successfully encrypts and then decrypts a message, one must obtain the original message. Thus was born the RSA algorithm, named for its inventors Rivest, Shamir, and Adleman.

The RSA algorithm functions in the following way. Anyone who wishes to receive an encrypted message chooses two prime numbers p and q. He or she then specifies numbers e and d with the property that the product ed is of the form $k \cdot (p-1)(q-1)+1$ where k may be any natural number. The numbers e and $n = pq$ form the public key. The recipient publishes these two numbers, but considers d to be a private key that is kept secret.

Anyone who wishes to send an encrypted message to this recipient encrypts the message m with the aid of the public key. This means the sender represents the message as a number m, calculates m^e and determines the remainder of m^e after division by n. This remainder is a number c that represents the encrypted text. The sender then sends the encrypted text c to the recipient.

The recipient applies the private key to decrypt the text. He or she calculates the remainder of c^d after division by n. Euler's theorem guarantees that the original message m results.

Why is the RSA algorithm a public key encryption system? One can convince oneself that the question of how difficult it is to calculate the private key using the public key is intimately related to the question of how difficult is it to factor the number n into its prime factors. In other words, one must choose prime numbers p and q such that no one can determine p and q from their product. The most important measure is to select large prime numbers for p and q. Today one chooses prime numbers with over 150 digits.

The current world record for the factorization of an RSA number $n = p \cdot q$ is the factorization of the number RSA-768. This number has 768 bits, which means it has 232 digits; it is the product of two prime numbers, each with 116 digits. This record was achieved in December 2009 by an international European team.

Chapter 2

Representations of numbers

2.1 How were numbers written in former times?

The need to record numbers is ancient. It probably originated in the practical requirement to represent a constant amount or its increase or decrease over time. If a goatherd wanted to check whether his herd contained the same number of animals on two consecutive days, then he made one notch for each animal and checked the next day whether the number of notches matched the number of goats.

Much later, potentates wished to be remembered after death. Numbers representing the year, amount of accumulated gold, etc., were thus carved in stone.

Other requirements gained importance later still: A number system should allow for a representation of large numbers, or better yet, of arbitrarily large numbers. Finally, a number system should allow for efficient computation.

The most elementary way of representing numbers is to make a corresponding number of tallies or notches. This method is ancient. Archeologists have found bones as old as 30,000 years with many notches. The Ishango bone, which was introduced in Chapter 1, shows the prime numbers 11, 13, 17, and 19 in the form of tallies:

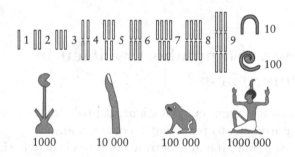

19   +   17   +   13   +   11   =   60

The Ishango bone

Soon early humans noticed that such lists of markings become confusing when recording even moderately large numbers. Therefore, in practically all cultures, the idea of grouping certain numbers became accepted, most often in sets of five and ten. This may mean leaving a small gap after each set of five notches, or it may mean indicating five-ness by a horizontal bar, as we do today. Or it might mean introducing new symbols for five and ten.

The Egyptian notation

The Egyptians recorded numbers systematically as early as 2000 B.C.: They had symbols for one, ten, one hundred, one thousand, ten thousand, one hundred thousand, and for one million. The Egyptians represented a number by writing down the appropriate number of copies of each individual symbol. They began on the left starting with the symbol of the highest value. This is an example of an additive system.

The Romans used the familiar system consisting of I (one), V (five), X (ten), L (fifty), C (one hundred), D (five hundred), and M (one thousand). Initially, this was also an additive system, so that for example the number 44 was written like this: XXXXIIII. Later the number 4 was no longer written as IIII but as IV, for brevity.

As compelling as the number systems of the Egyptians and the Romans may be at first glance, from a mathematical point of view they are both problematic.

The first problem is quite obvious: How did the Romans write 10,000? Perhaps by writing M ten times? And how did they represent one million? By writing M one thousand times? Of course the Romans did come up with solutions to these problems. For example, an M in a box meant 100,000. However, this does not truly solve the problem, it only delays it. The Romans (and the Egyptians) had to introduce new symbols for each new order of magnitude.

The second problem becomes apparent when one attempts to calculate with Roman numerals. Addition is quite all right: What is XVII plus LXI? That's easy: Write the number symbols consecutively, and then group them together. Thus XVII + LXI = XVIILXI = LXXVIII. Of course this only works if one uses a strictly additive system; as soon as abbreviations such as IV are used, even addition becomes challenging and prone to error.

Multiplication is even harder. What is XVII times LXI supposed to be? This calculation does not work at all in the Roman system; Roman numerals are not made for computing. The Romans carried out all calculations using an abacus, which is essentially a decimal place value system (see Section 2.2).

A completely different system was used by the Babylonians around 2000 B.C. in Mesopotamia, a region in today's Iraq. It is considered the most powerful number system of antiquity, far superior to any other system. It is what we now call a "place value system." Such systems are characterized by a limited supply of digits, which have different values depending on their "place." We are quite familiar with this: The one in the number 1,302 has a different value than the one in the number 2,310: In the first example, it is worth one thousand, in the second only ten.

The Babylonians did not use the decimal system we are used to, that is, a base 10 system. Instead they used base 60. The digits were the numbers $1, 2, \ldots, 59$. The last digit of a number was the ones digit, the second to last was the 60s digit, and in the third to last place, a one had a value of 3,600. The "sexagesimal number" 234 would be 7,384 in the modern base 10 system because $2 \times 3,600 + 3 \times 60 + 4 = 7,384$.

The Babylonians used an additive system to represent the digits, that is, the numbers $1, \ldots, 59$. A vertical bar meant one, and a wedge ($<$) represented ten. The digit 14 was therefore written as $<||||$.

The sexagesimal system performed outstandingly well when it came to calculations, so it was always used when many difficult calculations had to be performed, especially in astronomy. Even today, this system has left obvious traces: We divide one hour into 60 minutes, and one minute into 60 seconds. This means that the sexagesimal number 234 is the number of seconds in 2 hours, 3 minutes, and 4 seconds. The way we measure angles in degrees also has its origins in Babylonia.

By the way, the Mayans in Central America also used a place value system during the peak of their culture (until around A.D. 900), a base 20 system. The Mayans even had a concept of zero, and their symbol for it was a small seashell, arguably the most beautiful zero that has ever existed. The Mayan culture experienced a dramatic decline in the 10th century, and the culture did not continue. Thus the advanced mathematics developed by the Mayans had no impact on the further development of number systems.

2.2   The abacus and counting board

How did the Romans calculate? Undoubtedly, they *did* calculate. Without the elementary operations, one can neither build the Colosseum nor maintain an effective military, let alone establish a workable tax system. The Romans must have calculated! However, Roman numerals are terribly unsuitable for calculations. In principle, one can add, but multiplying is inconceivable. Roman numerals were almost exclusively used to record dates.

The Romans calculated using an abacus. The abacus was invented in China as early as 1000 B.C. It is the oldest calculating tool, and has been adopted worldwide: In addition to the "classic" Roman abacus, there are the Chinese Suanpan, the Japanese Soroban, and the Russian Stschoty. Some of these were used well into the 20th century.

The abacus is a decimal system in disguise. Its most elementary function is to represent a natural number. An abacus consists of several horizontal rods. The lowest rod represents the ones, the next one up the tens, followed by the hundreds, the thousands, and so on. The rods hold beads or pebbles, which are used to represent the value of the respective digit. Every rod has two halves. The beads on the left side represent one, the beads on the right, five. The value of a digit is determined by how many beads are moved to the center. The simplest version of the abacus has four beads to the left and one bead to the right of the center. This allows for the unique representation of the digits 1 through 9.

Two numbers are added by placing the first number on the abacus, followed by the second one. To compute the sum of 21 and 13, we place 21 on the abacus and add on 13. This means sliding 3 beads on the lowest rod to the middle, and one bead on the second lowest rod. Then we may read the result directly off the abacus: 34.

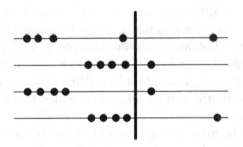

An abacus representing the number 1954

The calculation 21 + 34 is more challenging. In addition to the one bead on the bottom, we have to slide another four beads over—but there are only three beads there! Thus we slide these three beads to the middle and make a mental note that technically we have to slide another bead over. We know the bead on the right has a value that is one greater than the total value of the beads on the left. Therefore, we slide the bead on the right into the middle and the beads on the left back to the outside.

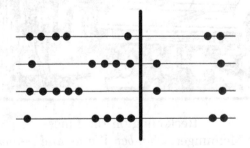

An abacus with carried numbers

Not only does this sound complicated, it really is complicated and also terribly prone to error. The abacus was thus developed further, with five beads to the left and two to the right of the middle on each rod. This allows for much better control of the carried numbers. In our example of 21 + 34, we add four beads to the one bead on the left, then replace these five beads with one bead on the right: We slide the five beads on the left to the outside, and at the same time move one bead on the right to the middle. Similarly, we replace two beads in the middle on the right by one bead on the left on the next rod up.

But even with this type of abacus it is difficult to calculate. One must learn how to do it, practice it, and remain focused. There is no way to retrieve a previous result, because the abacus does not provide a written record. For an observer, the pattern of the procedure is always the same: A bunch of numbers are read and represented on the abacus. Then a knowledgeable arithmetician spends some time calculating, and finally a result appears, which is read and converted to a Roman numeral.

The idea of the abacus was adopted in the Middle Ages in Europe, but instead of the abacus the Europeans used a counting board or a counting cloth. Sometimes the necessary lines were simply drawn onto a table top. This was known as "Reckoning on the Lines."

Reckoning on the Lines
(From Karl Menninger: *Number Words and Number Symbols,*
Göttingen, 1987)

Arithmeticians drew four horizontal lines. The lowest line represented the ones, the second lowest the tens, the third the hundreds, and the one on top the thousands. "Counting pennies," or simply pebbles, were placed on or between the lines. The number of pebbles determined the value of a digit; one pebble in the space between two lines (in a "spatium") counted five times as much as one pebble on the line below it. Often a vertical line divided the board in two halves so that two numbers could be represented at the same time. In the following picture we see the number 328 on the left, and 2,763 on the right.

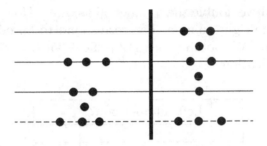

Two numbers represented on a counting board

How was this used for calculations? Adding numbers was, in principle, easy: Arithmeticians represented the summands on the left and on the right as shown above, then combined them as shown below on the left, and finally cleaned things up. In the following picture, we literally push the numbers from the above example together, then convert the interim result to the standard representation in the next two steps. The final result is the number 3,091.

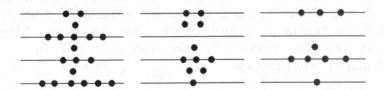

Addition and converting to the standard representation

The operations of doubling and halving were of particular importance. Doubling is easy, since it just means adding a number to itself. Halving is quite a mechanical procedure when using the "counting pennies": From a line with an even number of pebbles, remove half. If a lines starts with an odd number of pebbles, slide one into the space below, then remove half of those remaining. A pebble that starts in a space is replaced by two pebbles on the line below and one pebble in the space below. Odd numbers are trickier; in this case, a pebble slides into the space *below* the lowest line. Such a pebble is then omitted. This procedure is thus "halving with rounding down"; so 13 is halved to 6.

Doubling and halving are important because these operations can be used to multiply efficiently by only halving and doubling. To compute 83 times 56, we divide 83 into halves until we arrive at the number 1; and we double the number 56 as often. Now we add those numbers in the right column that correspond to an odd number in the left column. We obtain: $56 + 112 + 896 + 3584 = 4648$, the result of the multiplication.

	Left number odd?	
83	✓	56
41	✓	112
20		224
10		448
5	✓	896
2		1792
1	✓	3584

This procedure probably emerged in several places at different times. For example, the Egyptians and the Romans may have used it to multiply; it is also known as "Russian farmer's multiplication." In the end, this procedure is based on the binary system (see Section 2.5). One might deduce this when considering what happens when we multiply a power of two, for example $16 = 2^4$, by an arbitrary number x in this way:

	Left number odd?	
16		x
8		$2x$
4		$2 \cdot 2x = 4x$
2		$2 \cdot 4x = 8x$
1	✓	$2 \cdot 8x = 16x$

Since only the last row has an odd number in the left column, the result of the calculation is the entry in the last row.

2.3 The decimal system

> *There are few things on Earth that are perfect,*
> *and perfect customs which man*
> *has successfully created are fewer still.*
> *But he can pride himself on this:*
> *these new Indian numerals are indeed perfect.*
>
> Karl Menninger

Our numbers come from India. In particular, our zero was invented in India. Together with the other nine digits, it forms the basis for the decimal system. Place value systems were invented in several locations at different times, but the unrivaled decimal system used worldwide today has its origins in India.

The first documented Indian zero occurred in the year A.D. 786. On a stone tablet from Gwalior, a town about 500 km south of New Delhi, zero was used to represent the numbers 270 and 50. It appears so nonchalantly that one may safely assume that zero had previously been in use.

As nondescript as it may seem, zero is indispensable for the representation of numbers and computing. The Babylonians had a place value system over 2,000 years before the Indians, but they did not have a zero! If we did not have a zero, then we would not be able to write the number 205 as we do. Instead we would have to write 2, leave a gap for the "empty" tens digit, and then write 5. Clearly, this notation inevitably leads to misunderstandings: One person might interpret "2 5" as 25, assuming that the spacing is just a bit off. Someone else might think that there is a gap at the end and call it 250.

The Indians didn't just use zero, represented by a small circle, to write numbers. Instead, they considered it a number in its own right. For example, they noticed that a number minus itself equals zero, and that one may add zero to any number without changing its value.

Starting in India, zero, and with it the decimal system, began its victory lap around the world. The new number system gained speed with the rapid spread of Islam. During its expansion after Mohammed's death (A.D. 632), Islam eagerly picked up, processed, and developed the cultures of conquered countries, and thus preserved and harnessed them for modern times.

Let us highlight two mile-markers along the way. The most distinguished mathematician at the time, Abu Dscha'far Muḥammad ibn Mūsā al-Khwārizmī, lived in Baghdad from around A.D. 780 to 850. His "Book of Addition and Subtraction According to the Hindu Calculation" is the first book outside of India to describe the decimal system, including the number zero. It was the first step toward worldwide recognition of the Indian system. The book itself was only passed on to us in a Latin translation. The author's name al-Khwārizmī was Latinized as "al-gorismi." It developed into the word "algorithm," which generally means a computational procedure following a set of fixed rules. Moreover, the title al-dschabr of another one of al-Khwārizmī's books developed into the word "algebra."

Fibonacci (actually, Leonardo of Pisa, c. 1180–1250) is considered the first "modern" European mathematician. In his 1202 *Book of Calculation (liber abaci)*, he introduces the new Indian system and its advantages. In particular, Fibonacci writes: "The nine Indian digits are 9 8 7 6 5 4 3 2 1. With these nine digits, and with the symbol 0, which the Arabs call zephirum, any number may be represented." This mathematical theorem proves to be of immense value in the plethora of exercises that follows in his book.

However, the number zero and the Indian decimal system were not immediately accepted; on the contrary, they had to prevail in a fight against established traditions over hundreds of years. One argument against them involved employment: With the new decimal system, pretty much anyone could calculate, whereas with the old system, special skills were required. Another argument against the new system was the claim that the new digits were easy to forge: One brush stroke sufficed to change a 0 into a 6 or 9. Not until the 16th century was there a general demand for learning and using the new method. This is documented by a large number of books on arithmetic, among them the bestseller *Rechenung auff der linihen und federn* (1522) by Adam Ries. In addition to computing with a counting board (the "linihen" being the lines), Ries describes how to perform written calculations (with "federn" being feather pens) using the Hindu-Arabic digits. Subsequently there has been no stopping the triumph of the decimal system.

2.4 Divisibility rules

"A number is even when its ones digit is even." This theorem is often regurgitated without much thinking. Some even think this is the definition of an even number—but this is wrong. The definition of an even number is that it is divisible by 2 without leaving a remainder.

The fact that we can read this property off the last digit of a number is a mathematical theorem, and like every mathematical theorem is something of a miracle. Suppose we want to know whether the number 782,573,728,764,104 is even. By definition, we have to divide this monster by 2 and check whether the remainder is 0. The theorem shows us a much easier method: It suffices to take a closer look at the last digit, 4, and we know at first glance whether this huge number is even. This is a "David and Goliath theorem": Based on a tiny observation (the last digit), we knocked out a gigantic number!

We can convince ourselves of the truth of this theorem in several ways. The most illustrative method uses the figurate numbers from Chapter 1. We write the number as a rectangle of height 10. Of course in general, this will be an imperfect rectangular number; it probably will not actually make a rectangle, and there is a remainder that is less than ten. This remainder is the ones digit. Let us consider the number 54. It consists of 5 tens and 4 ones. Written as a figurate number, it looks like this:

54 as a figurate number

The first 5 columns consist of ten dots each. Since 10 is an even number, the number containing all but the last column must be even. Whether the overall number is even therefore only depends on the last column. If that number is even, then the overall number is even, and otherwise it is odd.

This proof may also be expressed in a symbolic way. To check whether a natural number n is even, we write it as $n = 10a + b$.

In the case of $n = 54$, we write $n = 10 \times 5 + 4$. Here, the number b represents the ones digit. We proceed in a similar manner: $10a$ is always an even number, since 10 is even, and so is every multiple of 10, in particular $10a$. Now we apply the old rule of the Pythagoreans that "even plus even is even." In our example this means: If we take the even number $10a$ and add an even number b to it, then we obtain another even number. Therefore $n = 10a + b$ is even. On the other hand, "odd plus even is odd." So if we add an odd number b to the even number $10a$, then the sum $n = 10a + b$ is odd. In summary, this means: $n = 10a + b$ is even if and only if the last digit b is even.

Even more exciting than divisibility by 2 is divisibility by 9. The theorem in question is another example of a "David and Goliath theorem." To decide whether the number 123,456,789 is divisible by 9, all we have to do is add its digits, or in other words, find its "checksum"; the checksum of 123,456,789 is $1 + 2 + 3 + 4 + 5 + 6 + 7 + 8 + 9 = 45$. Since 45 is divisible by 9, the monster number above is also divisible by 9.

Every number is either even or odd. In other words, with divisibility by 2, there are only two cases. When divided by 2, a natural number leaves a remainder of either 0 or 1.

But divisibility by 9 has even greater potential! Of course one might say: Either a number is divisible by 9, or not. Alternatively we might take a closer look and determine not only "remainder 0" and "other remainder," but distinguish between all possible remainders: When divided by 9, a natural number leaves a remainder of 0, 1, 2, 3, 4, 5, 6, 7, or 8. This is called the "nines remainder" of a number. For example, the number 1,024 has a nines remainder of 7, because $1,024 = 9 \times 113 + 7$.

It is easier to come to this conclusion by considering the checksum. The checksum of 1,024 is $1 + 0 + 2 + 4 = 7$. The nines remainder of 7 is, of course, 7—and this is also the nines remainder of our original number.

This surprising mathematical fact holds true in general: A natural number leaves the same remainder as its checksum when divided by 9. This leads us to a generalization of the nines rule. Since a number is divisible by 9 if and only if its nines remainder is 0, one may rewrite the nines rule as:

The nines remainder of a number is 0 if and only if the nines remainder of its checksum is 0.

This mathematical fact provides a framework for a quite intriguing magic trick. Ask someone to think of an arbitrary five-digit number and write it on a piece of paper, so that you are unable to see the number. Then instruct them to write the digits of the number down in a different order. Then they should subtract the smaller of the two numbers from the larger one. Finally, ask them to choose and circle one digit of the result. If the result contains a zero as a digit, then it may not be circled. You can always say that the number zero is already a circle itself. Now ask them to read the other digits to you—and you will be able to name the circled digit immediately.

The calculation may for example look like this: The original number was 54,831; the permutation was 38,415. Then the difference is 16,416. The digit in the middle was circled, and they read the digits 1, 6, 1, 6. All you have to do is add the digits that were read back to you: $1 + 6 + 1 + 6 = 14$, and find the difference to the next nines number. Here, we have to add 4, and that turns out to be the circled number.

But wait, there is more! The nines remainder has an application of historical importance, the "nines check," which Adam Ries held in high regard. The nines remainder allows us to check whether a complicated calculation was carried out correctly. It is an indicator of the inner harmony of a calculation.

Roughly, the rule may be expressed like this: If a calculation was carried out correctly, then it must also be correct when carried out with the respective nines remainders. This is because the nines remainder gets along famously with the elementary operations: If we want to calculate the nines remainder of a sum $a + b$, then we may find the nines remainders of the summands a and b and then add them. Notice that our magic trick above depends on the fact that the nines remainder cooperates with subtraction.

In order to check whether the sum $247 + 354 = 601$ was calculated correctly, we add the nines remainders 4 and 3 of the numbers 247 and 354 and compare this to the nines remainder of the sum 601. If these numbers are not the same, then we know for sure that a mistake was made in the calculation; if the two results are the same, then the calculation was probably done correctly.

The same thing goes for multiplication: The nines remainder of a product is the product of the nines remainders. For example, the nines remainder of 19 times 31 is the same as the nines remainder of 19 (i.e., 1) times the nines remainder of 31 (i.e., 4); the result is 4.

The product of 19 and 31 is 589; since this also has a nines remainder of 4, the calculation was probably carried out correctly.

Adam Ries describes the nines check as a schematic procedure. To verify the correctness of the calculation $7,869 + 8,796 = 16,665$, we begin by drawing a cross in the form of a big X. We write the nines remainder of the first number, that is, of 7,869, on the left; it is 3. The nines remainder of the second number, that is, of 8,796, goes on the right; it is also 3. We add the two nines remainders and put the result on top, i.e., $3 + 3 = 6$. If the result is greater than or equal to nine, then we subtract 9 until we obtain a number between 0 and 8. Finally, we put on the bottom the nines remainder of the sum $7,869 + 8,796$, that is, the nines remainder of 16,665: It is 6. If the same number appears on top and on the bottom, then everything checks out; if the numbers are different, there is a mistake in the calculation.

2.5   Binary numbers

The Pythagoreans discovered the distinction between even and odd numbers. From there originated the description of the world through polar opposites: above and below, left and right, before and after, day and night, old and young, life and death. Mathematicians express this, very matter-of-factly, with plus and minus, or simply with 0 and 1. Not only is it possible to identify opposites this way, but moreover, all numbers may be written in terms of 0s and 1s, something that the great Gottfried Wilhelm Leibniz published in 1703.

His idea was to develop a place value system using only 0 and 1. This is known as the *binary system*. In principle, things work just like they do in the decimal system: The value of a digit depends on its place. In the decimal system, a 1 as the last digit just means one, in the second to last place it has a value of ten, in the third to last place one hundred, and so on.

In the binary system, a 1 in the second to last place has a value of two; the number two is thus written as 10. The binary number 11 means three; since the 1 in the last place has a value of one, and the 1 in the second to last place has a value of two, and these add up to three.

Does that make sense? Let's take a step beyond this. A 1 in the third to last place has a value of four (since 2 times 2 is 4). This means that the number four is written as 100, the number five 101, the number six 110, and finally the number seven as 111. Clearly, since 111 means: 1 one, 1 two, and 1 four, together this adds up to seven. The first few numbers are therefore written in binary as follows: 1, 10, 11, 100, 101, 110, 111, 1000, ...

Decimal number	Binary number
0	0
1	1
2	10
3	11
4	100
5	101
6	110
7	111
8	1000
9	1001
10	1010
11	1011
12	1100
13	1101
14	1110
15	1111

The binary numbers from 0 to 15

This is the representation of numbers used by modern computers. All data, regardless of whether they are text, audio, or video data, are represented by numbers, and numbers are represented by zeros and ones, i.e., bits. What we take for granted today is the original vision of the American mathematician Claude E. Shannon (1916–2001), who published an article in 1948, in which he described his insights: Any kind of information may be represented in the form of "bits" ("binary digits"). Without Shannon's idea, the development of information technology would most likely have taken a different path.

Before Shannon, Leibniz had envisioned a binary computer. However, he did not pursue his idea of a "Machina Arithmeticae Dyadicae," because his computing machine based on the decimal system had already confronted him with almost unsolvable problems.

Leibniz had predecessors: Representations of numbers using only two symbols were found in India and China; whether the symbols in question represent numbers or whether they are only a systematic sequence of combinatoric patterns remains unclear. In Europe, too, binary number systems had been published a few decades before Leibniz. However, it would be fair to say that Leibniz was the first to recognize the enormous potential of binary numbers. This potential lies in the fact that it is extremely easy to calculate with them. For example, the entire "binary multiplication table" reduces to the one trivial equation, $1 \times 1 = 1$.

Leibniz writes: "Addition of numbers by this method is so simple that they can be added as quickly as they can be written." Today, we say that the "cost of addition is linear." This means that the time required for adding two numbers is just a constant multiple of the time required for writing the numbers.

2.6   Applications: Barcodes

Errare humanum est.

We read and write numbers, and we make mistakes. How often do we overlook a digit? How easy is it to read 8 instead of 3? How often do we say "twenty-three" but write 32?

There are a number of traditional methods for dealing with such errors. If we did not understand a word, then we ask for it to be repeated. Sometimes numbers are written as words, and difficult words may be spelled out. All of these methods have something in common: The actual information is augmented by additional elements. A word is repeated, or in addition to a digit, a word is transmitted; a word is verified by the sequence of its letters. In short, we add redundancy. Often, a lot of redundancy is added, and the number of symbols to be transmitted at least doubles. Mathematical procedures for error detection are based on the same idea, but they make do with minimal redundancy, that is with only one additional symbol.

This idea is best illustrated by a small example. Suppose we send the message 123456 and would like to protect it from erroneous reading or writing. We may add a check digit to the original message 123456 in such a way that the total sum of the digits is a multiple of ten. Since $1 + 2 + 3 + 4 + 5 + 6 = 21$, the check digit is 9, and the complete message reads 1234569. We call such a procedure, or the set of messages created in this way, a "code."

The receiver of the coded message reads some sequence of digits. In order to check whether the message was transmitted correctly or not, they compute the check digit. Or even easier: They determine whether the checksum is correct. If an error occurred, for example if instead of 3 an 8 was transmitted, then the received message is 1284569. Since the checksum of this number is 35, the message will not be accepted.

In general, these types of codes will detect an incorrect digit, in the sense that the checksum is no longer a multiple of ten. However, if errors occur in two or more places, then the effects may cancel out. We say that such a code "detects one error."

For many purposes these types of codes are sufficient. However, they do not catch transposition errors: The message 1324569 is accepted as much as the original message 1234569.

In order to detect the latter type of error, one has to use another version of the checksum, which distinguishes the neighboring digits. This may be realized by "weighting" the digits differently. Often the first, third, fifth, etc., digits are added, while the second, fourth, and sixth digits are multiplied by 3 before they are added to the other digits. In our example, we would compute $1 + 3 \times 2 + 3 + 3 \times 4 + 5 + 3 \times 6$, which is 45. Again, we complete this number to the next multiple of ten. The complete message is then 1234565.

This code will detect single errors and most transposition errors. For example, the sequence 1324565 leads to a checksum of $1 + 3 \times 3 + 2 + 3 \times 4 + 5 + 3 \times 6 + 5 = 52$. Since this is not a multiple of ten, the message will be rejected.

The weighting of $1, 3, 1, 3, 1, \ldots, 3$ is the basis for the "European Article Number (EAN)," which is known worldwide as a "barcode." The bars are simply the translation of the number below into a machine-readable form. The barcode is directly useful to us as consumers: When the cashier scans an item, then the check digit is computed automatically. The beep means that everything is alright; and we can be sure that the correct item was detected and we only pay for this particular item.

These types of error correcting codes are but the beginning of an extensive theory, which has numerous practical applications. In practically every transmission of data, error correcting codes are used. No cell phone or CD can work properly without them. In practically every transmission of data, error correcting codes are working behind the scenes.

Chapter 3

Rational and irrational

There are no secrets in the numbers,
alas the fractions hold a great one.

Goethe, *Urfaust*

3.1 Fractions

Describing the world with natural numbers was an advance in knowledge whose extent we are barely able to imagine today: It was possible to illustrate things and collections of things through numbers, and one could distinguish between them and compare them. However, the attempt to capture more and more phenomena through numbers posed a fundamental problem: One had to break whole numbers up into parts!

The more or less exact process of halving was surly a well known technique early on. One had to cut a piece of meat into halves, or distribute a certain number of nuts among two people, or divide a piece of land into two equal pieces. Dividing things into more than two pieces was also an everyday necessity.

Today we take fractions for granted: We speak of a "half pint," a "quarter pound," the "third third" of a hockey game, a "seven and a half ton" truck, "quarter to noon," and so on.

In short, the problem was to divide an object or a multitude of objects into a given number of equal parts. The easier part of the problem was the actual dividing of the objects; the harder part was

41

the task of representing this process by numbers and thus calculating and controlling it.

It was no longer possible to work only with the natural numbers. Instead it became necessary to represent dividing a certain number of objects in the language of numbers. It was not sufficient to work with whole, unbroken numbers—they had to be divided, that is broken up into smaller pieces, or "fractions."

What was easy to accomplish in practice turned out to be a huge problem theoretically. In the following we will outline some attempts to solve this problem.

First solution: Avoid fractions

The Mesopotamians used a base 60 place value system as early as 2000 B.C. (see Chapter 2). We may ask ourselves why the Mesopotamians chose 60 of all numbers as a base. The reason is probably that 60 is divisible by surprisingly many numbers, more than any smaller number: 60 is divisible by 2, 3, 4, 5, 6 ,10, 12, 15, 20, and 30, without leaving a remainder. This is why the Babylonians could represent the fractions 1/2, 1/3, 1/4, 1/5, 1/6, 1/10, 1/12, 1/15, 1/20, and 1/30 very easily.

It is like using decimals. The divisors of 10 are 2 and 5, and therefore the fractions with these denominators are easily represented: $1/2 = 0.5$, and $1/5 = 0.2$. All other fractions create problems, for example 1/3, 1/6, and 1/7. The Babylonians were happy with the fractions and calculated with them in a virtuosic way. They could solve all practical and many theoretical problems.

Second solution: Smaller units

In principle, this is the same idea that Babylonian fractions are based on, but it becomes particularly apparent in monetary systems. The Romans had a copper coin, called an *as*, that weighed 12 ounces. It was possible to express fractions of an *as* by an appropriate number of ounces: 1/12 of an *as* is one ounce, 3/4 of an *as* is 9 ounces, and so on. In reverse, "semi" (half) means 6 ounces, "quadrans" (quarter) means 3 ounces, and "sextans" (sixth) means 2 ounces. This idea has been preserved to the present day. Until 1971 there was a coin in England that was called both "sixpence" and "half-shilling."

Third solution: Unit fractions

As early as 2000 B.C. the Egyptians came up with a radical, systematic approach. For every natural number n, they defined the unit fraction $1/n$ by writing the hieroglyph for "mouth", a thin oval, over the hieroglyph for the number. We denote this by writing a bar: $\overline{n} = 1/n$. Apart from the fractions $2/3$ and $3/4$, for which they had special characters, the Egyptians only used unit fractions, that is, fractions such as $\overline{2}$ $(=1/2)$, $\overline{3}$ $(=1/3)$, and $\overline{7}$ $(=1/7)$.

Of course, often many other fractions occurred; these had to be represented as sums of unit fractions. For the Egyptians, the "fraction" $2/5$ was not the final result of a calculation, but an interim result that must be written as a sum of distinct unit fractions, that is, $2/5 = 1/3 + 1/15$. This was the only acceptable result. The "trivial" representation, for example rewriting $14/25$ as the sum of fourteen fractions of the form $1/25$, was unacceptable. One was obliged to write a fraction as the sum of distinct unit fractions. For example: $14/25 = 1/2 + 1/17 + 1/850$. (Verify this result!) We see at first glance that this is difficult. The Egyptians had nonetheless mastered complex techniques for writing fractions in this way.

The famous Rhind papyrus, which together with the Moscow papyrus contains everything we know about Egyptian mathematics, includes a table of fractions of the form $2/n$ and their representation in terms of distinct unit fractions: $2/3 = \overline{2} + \overline{6}$, $2/5 = \overline{3} + \overline{15}$, $2/7 = \overline{4} + \overline{28}, \ldots$. In general the equation $2/(2n+1) = \overline{n+1} + \overline{(n+1)(2n+1)}$ holds, which can easily be verified by finding a common denominator.

The Egyptians considered fractions as numbers in their own right—not just as ratios of whole numbers. However, the representation of fractions as the sum of distinct unit fractions was a notational dilemma. The Egyptians had to resolve this issue, and they were only able to calculate with unit fractions in basic ways; essentially they could only add and subtract them.

Fourth solution: Indian fractions

The representation of fractions that we are familiar with today originated in India and has been in use there since c. 600 B.C. The Indians wrote fractions just like we do, with the numerator on top and the denominator on the bottom. However, they abstained from using a fraction line.

The price Indian mathematicians paid for representing various different fractions as a single number was the lack of uniqueness of fractions. We know that $1/2 = 2/4 = 3/6 = 16/32$. If a fraction may be transformed into another by canceling or expanding, then the two fractions represent the same number. A rational number may therefore be represented by an infinite number of fractions! The ambiguity of the representation, that is, the divergence of the represented (the rational number) and its representation (the fraction) is unfamiliar and initially tends to cause difficulties. However, one may also view the diversity of the representations of a certain rational number as a metaphor for freedom: Among infinitely many fractions, we may choose the one that works best for a given purpose. We take advantage of this when we work with fractions (see Section 3.3).

During the middle ages practical calculations with fractions became the norm in Islamic as well as Western European mathematics. In scores of books written by arithmeticians, extensive instructions were given along with examples for performing calculations with fractions. Although there was no theory of rational numbers, people got used to dealing with them and were unconsciously convinced that they "exist".

3.2   Ratios

The Greeks considered numbers as the result of a counting process. In book VII of the *Elements*, Euclid writes: "A number is a quantity composed of units." Expressed in today's language, the Greeks only knew the natural numbers. Basically the idea of a "fraction" contradicts the Greek notion of what a number is fundamentally. However, the Greeks also needed a method that allowed them to handle parts of whole numbers. The idea was to use ratios of natural numbers.

The Pythagoreans made a discovery that was to have a decisive impact on mathematics. Their discovery connects the realm of numbers with the realm of sounds. More precisely, the Pythagoreans knew how to describe every sound produced by two notes, that is, every musical interval, as the ratio of two natural numbers.

The monochord is a musical instrument used by the Pythagoreans not only to play music, but also for experimental purposes. This instrument has only a single string. Plucking it generates a tone. One may hold the string down at a point, just as with a guitar or violin string, and then pluck one side or the other of the string. This typically results in two different pitches; they are only the same if the string is halved. Dividing the string according to the ratio 2 : 1, that is, "two thirds to one third," we hear two different notes that form precisely one octave, the purest sound of any musical interval.

Dividing the string according to the ratio 3 : 2 produces an interval known as a "perfect fifth"—also a very pure sound. And so on. By way of experiment, the Pythagoreans discovered the following surprising connection: The purer the sound, the simpler the ratio of the chord length. And conversely, the more complicated the ratio (for example 9 : 8), the more complex the resulting sound.

The conjunction of such different areas as tones and numbers, more precisely of pitches and ratios of numbers, must have been a revolutionary insight for Pythagoras and his students. From then on they were convinced that the whole world could be described by numbers. They expressed this fittingly in their motto: "Everything is number!" The Pythagoreans were convinced that it was possible to express literally everything through natural numbers and their ratios.

By then ratios or "proportions" were used to describe things for which the concept of natural numbers was insufficient. We would simply identify a ratio of 2 : 3 with the rational number 2/3. However, the Greeks did not make the leap to considering a ratio itself as a number, that is, to regarding a ratio as an object in itself instead of just as a relation between two objects. However, the power of the concept of ratios was soon to be revealed.

Multiplying quantities, and in particular line segments, is easy: Given a line segment, we can use a ruler and a compass to construct a line segment that is twice as long, three times as long, and so on.

Dividing a line segment into several equal parts is much harder. We know how to construct the midpoint of a line segment; in other words, we know how to halve the line segment. (Using a compass, we construct two points on the perpendicular through the midpoint and draw a line through these two points. Then the point of intersection of this line with the original line segment is its midpoint.) Iterating this process will result in four or eight equal parts. If we want to divide the line segment into three, five, or seven equal parts, however, we have to use a different method.

A few hundred years later, in Euclid's book we find a totally unexciting theorem saying that every line segment may be divided into n equal parts. In order to divide a line segment with end points A and B into three equal parts, we use an auxiliary line AC_1. On this line, we mark the points C_2 and C_3 such that the distance between two consecutive points is equal to the distance between A and C_1 (see the figure below). Now we draw a line through B and C_3, and we draw two lines parallel to BC_3 through C_1 and C_2. These parallel lines intersect with AB at the points B_1 and B_2, which divide the line segment into three equal parts.

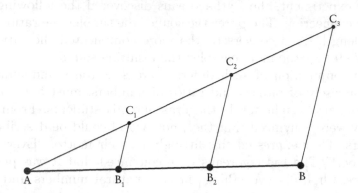

Dividing a line segment into three equal parts

In a similar fashion we may also construct an arbitrary fraction of a line segment: In order to find 4/7 of a line segment, we first construct 1/7 of the line segment and then quadruple the result.

With this construction, the Greeks accomplished a decisive step: They were now able to measure to arbitrary accuracy, and they could approximate any quantity to arbitrary accuracy using ratios. That is, the ratios are "dense."

It was an important problem in Greek mathematics to find a common measure for two given line segments. Euclid considered a "long" line segment AB and a "short" line segment CD. We say that the line segment CD "fits into" the line segment AB, if we can remove a line segment as long as CD from AB a certain number of times without leaving a remainder. In other words: CD "fits into" AB if the length of AB is an integer multiple of the length of CD.

If the line segment CD does not fit into AB, then we may remove line segments as long as CD from AB as many times as possible—but there will be a remainder. Let us call this small line segment EF.

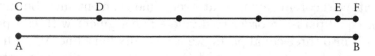

Measuring a line

Now we may ask whether EF fits into CD. Then EF also fits into AB, and we have found a common measure for AB and CD.

If EF does not divide CD, then one may proceed as follows: Remove EF as many times as possible from CD, and let the tiny remainder be denoted by GH. If this line segment divides EF, then it also divides CD and thus AB—and again we have found a common measure.

This procedure is called "measuring a line" because short line segments are repeatedly removed in order to "measure" a line segment. As Euclid describes it (*Elements*, VII, 2): "If CD does not measure AB, then, when we continually subtract the lesser from the greater of the numbers AB and CD, some number will be left which measures the one before it."

This expresses the idea that there is a common measure for every pair of line segments. Worded more cautiously: We call such line segments "commensurable"—and this suggests that there may be line segments that are not commensurable, that is, *in*commensurable. We may restate the paragraph from Euclid's *Elements* as follows: If the process terminates, that is, if at some point two consecutive line segments emerge such that the shorter one divides the longer, then the shorter line segment is a common measure of all interim line segments, and thus for the original line segments AB and CD. Today we call this procedure the "Euclidean algorithm"; we use it to determine the greatest common divisor of a pair of natural numbers.

For example, to determine the greatest common divisor of 101 and 35, we may proceed as follows: Let us imagine a line segment of length 101 and remove from it a line segment of length 35 as many times as possible (in this case, that is twice). The remainder is a line segment of length 31.

In a second step, we take the line segment of length 35 and remove a line segment of length 31; we are left with a remainder of 4. Now remove a line segment of length 4 from the line segment of length 31 seven times. The remainder is of length 3. Upon removing a line segment of length 3 from the line segment of length 4, we obtain a remainder of length 1. Since the latter divides the line segment of length 3, we have found a common measure of 1 for all the line segments we considered, and in particular for the line segments of length 101 and 35.

Today we write these geometric considerations as follows:

$$101 = 2 \times 35 + 31,$$
$$35 = 1 \times 31 + 4,$$
$$31 = 7 \times 4 + 3,$$
$$4 = 1 \times 3 + 1,$$
$$3 = 3 \times 1 + 0.$$

Using the Euclidean algorithm, we may easily determine the greatest common divisor of two very large numbers, without knowing their prime factorization. This technique plays a key role in many applications, including modern cryptography (see Chapter 1).

3.3   Rational numbers

Natura non facit saltus.

One can introduce fractions in several different ways: as portions, as fragments, or as proportions. In contemporary school books, one sees an assortment of models for fractions: pieces of cake or pizza, parts of rectangles and line segments.

We've become accustomed to referring to fractions as "rational numbers." This term comes from the Latin word *ratio*, which among other things means proportion. In the literal sense "rational numbers" are ratios, but they are the same as fractions.

Whole numbers are also fractions, because the number 5 for example can be written as $\frac{5}{1}$. This means the set of whole numbers is included in the set of fractions.

Calculations work wonderfully with rational numbers; for computation they leave hardly a wish unfulfilled. One can add them, subtract them, multiply and divide. These operations work as well as one might ask.

Adding and subtracting rational numbers is particularly easy, at least if we imagine a rational number as a distance, a "piece of cake," or any other suitable quantity. In order to determine $\frac{4}{7} + \frac{3}{8}$, we consider a distance of length $\frac{4}{7}$ and a distance of length $\frac{3}{8}$ and join the two pieces together. The result is a distance of length $\frac{4}{7} + \frac{3}{8}$.

This task is only difficult if we want to display the result as a fraction. First we set ourselves an easier task: If two fractions have the same denominator, then addition is easy. One fifth plus three fifths is equal to four fifths. We act as if a "fifth" refers to part of an object such as an apple or a pizza. One fifth of a pizza plus three fifths of a pizza is, naturally, four fifths of a pizza.

How does one calculate $\frac{4}{7} + \frac{3}{8}$? We apply a wonderful characteristic of rational numbers, namely that the same rational number can be represented by different fractions. It is clear that $\frac{2}{4}$ is the same as $\frac{1}{2}$ and that $\frac{3000}{8000}$ is the same as $\frac{3}{8}$. In other words, we can expand a fraction by multiplying its numerator and denominator by the same number—and the rational number does not change when we expand the fraction.

When adding two fractions we expand both so that they have the same denominator. At school we learned to find the "common denominator." In our example we multiply numerator and denominator by 8 to get $\frac{4}{7} = \frac{32}{56}$ and we multiply numerator and denominator by 7 to get $\frac{3}{8} = \frac{21}{56}$. Because these fractions have the same denominator, we can determine the sum: $\frac{32}{56} + \frac{21}{56} = \frac{53}{56}$.

When adding fractions we may be very flexible. We may reverse the order: $\frac{1}{2} + \frac{1}{3}$ is the same as $\frac{1}{3} + \frac{1}{2}$. If we are adding more than two fractions, then we may choose which plus sign to evaluate first. For example, if we wish to determine $\frac{1}{3} + \frac{1}{4} + \frac{1}{8}$, we can first calculate $\frac{1}{3} + \frac{1}{4}$ (which gives $\frac{7}{12}$) and then add $\frac{1}{8}$. This is $\frac{14}{24} + \frac{3}{24}$, which is $\frac{17}{24}$. Alternatively, we can evaluate the second plus sign first by calculating $\frac{1}{4} + \frac{1}{8} = \frac{3}{8}$, and then adding $\frac{1}{3}$. The result is $\frac{8}{24} + \frac{9}{24} = \frac{17}{24}$. The order of the calculations doesn't matter—the result is the same!

These characteristics of addition are called the commutative and associative laws. I find the term "law" deceptive, because we tend to think that laws place constraints, whereas here we are concerned with the possibilities of freedom. You may choose the order in which you add certain fractions, and I can choose a different order—and we will both get the same result!

Products and quotients of fractions are more difficult to imagine. The Greeks always thought of numbers as lengths. Then it is clear that a length times a length is an area. The product of a length of 4 units and a length of 5 units is a rectangle of 20 units in area.

That sounds good, but obscures a fatal flaw. The product of two lengths is an area, which is in principle already very different from a length.

Furthermore, the product of three lengths is a volume, and the product of four lengths is impossible to envision.

During division this concept leads to more difficulties. What is 12 divided by 3? In order to solve this problem, we can interpret it in different ways:

1. Consider a rectangle 12 units in area. If one side is 3 units in length, how long is the other side? The answer is simple: 4 units. Note: Here 12 is an area and 4 is a length!

2. Consider a length of 12 units. How many times does a length of 3 units fit into a length of 12? The answer is simple: 4 times. Note: Now 12 is a length and 4 is a number!

René Descartes (1596–1650), inventor of analytical geometry, or geometry in which calculations are made within a coordinate system, achieved an intellectual breakthrough. About multiplication he wrote: "Let AB be the unit, and let us multiply BD and BC. I have only to join the points A and C and then draw DE parallel to CA. Then BE is the product of the multiplication."

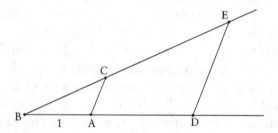

Descartes' method for multiplication of lengths

That is, in order to multiply the lengths BC and AD, one chooses the point A such that BA has length 1. One then constructs the point E as described by Descartes. Then

$$\frac{BE}{BC} = \frac{BD}{BA}.$$

If BA = 1, then it follows that BE = BD · BC. In other words, a length times a length is a length! Now we can multiply arbitrarily many numbers and we will always obtain a length!

This insight contributed a great deal to the abstract definition of a number. If both the factors and the product of the factors are lengths, then it is just a short step to forget lengths and only think about numbers.

The rules for multiplication and division of fractions are simple to write down, but they are not simple to understand. It is easy to calculate $\frac{2}{3} \cdot \frac{1}{2}$. One multiplies the numerator and the denominator, and thereby obtains the product $\frac{2 \cdot 1}{3 \cdot 2}$, which is $\frac{2}{6}$, or $\frac{1}{3}$.

How can this be explained? A careful verbal formulation will help. The task of computing $\frac{2}{3} \cdot 6$ can be described as "$\frac{2}{3}$ of 6," and it is clear that the result is 4. Analogously, the task $\frac{2}{3} \cdot \frac{1}{2}$ is in words "$\frac{2}{3}$ of $\frac{1}{2}$," and it is clear that the result is $\frac{2}{6}$. (One divides $\frac{1}{2}$ into three equal parts, or three sixths. Then two thirds of $1/2$ is two of these three parts: two sixths.)

Division by a fraction is more difficult to imagine. Let us consider the task $3 \div \frac{1}{2}$. The question that aptly describes this riddle is: How many times will $\frac{1}{2}$ fit into 3? This is easy to answer: $\frac{1}{2}$ fits 6 times into 3. Thus $3 \div \frac{1}{2} = 6$.

We can calculate $\frac{2}{3} \div \frac{1}{2}$ in the same way. How many times does $\frac{1}{2}$ fit into $\frac{2}{3}$? Equivalently, we can ask: How many times does $\frac{3}{6}$ fit into $\frac{4}{6}$? Analogously, how often does 3 fit into 4? The answer: precisely $\frac{4}{3}$ times.

Division can be traced back to multiplication: To divide by a fraction is equivalent to multiplying by its reciprocal. If we wish to divide by $\frac{2}{3}$, we may as well multiply by its reciprocal, $\frac{3}{2}$. For example:

$$\frac{1}{2} \div \frac{2}{3} = \frac{1}{2} \cdot \frac{3}{2} = \frac{1 \cdot 3}{2 \cdot 2} = \frac{3}{4}.$$

Practical calculation with fractions was widely accepted in the Middle Ages. It was accepted de facto that $\frac{2}{3}$ is not only a relationship between whole numbers, but also a new sort of number.

Mathematical theorists addressed this practical knowledge much later. In the middle of the 19th century, Bernard Bolzano (1781–1848) in his *Pure Theory of Numbers* developed a "theory of number systems that are closed with respect to the four basic arithmetic operations." Bolzano viewed a number system in terms of the operations: A sensible set of numbers must include the natural numbers and must be of such a nature that one can add, subtract, multiply, and divide every pair of numbers and in every case obtain a number as the result. (The sole exception is that one is not allowed to divide by zero.) Bolzano said that one must first define numbers (plural!) in order to explain what a number (singular!) is. His main criterion for calling the elements of this set "numbers" was that one can then calculate without difficulty, which meant add, subtract, multiply, and divide. Mathematicians call such a structure a "field," and speak of, for example, the "field of rational numbers."

Bolzano's approach represents a purely theoretical view of numbers. In practice, rational numbers were introduced because we need to take ever more precise measurements. The rational numbers achieved this goal almost perfectly.

Between every two rational numbers lies a third rational number— for example, their average, or their "arithmetic mean." Between 0 and 1 lies $\frac{1}{2}$. Between 0 and $\frac{1}{2}$ lies $\frac{1}{4}$; between 0 and $\frac{1}{4}$ lies $\frac{1}{8}$, and so on.

This observation has dramatic consequences:

- Between every two different rational numbers lie infinitely many rational numbers.

- Rational numbers lie arbitrarily close together. For example, there are rational numbers that lie arbitrarily densely around the number zero, for numbers of the form $\frac{1}{n}$ approach zero arbitrarily closely.

The image of the number line changed with the introduction of rational numbers: The number line is strewn with rational numbers. In every tiny interval lies not just one, but infinitely many rational numbers. From this perspective it is hard to imagine that one might need even more numbers or that more numbers might exist.

The second answer to the question, "What exactly is a number?" could be as follows: If we wish to divide numbers arbitrarily—that is, if we wish to apply the four arithmetic operations without constraint (with the exception of division by zero)—then the rational numbers are the right numbers.

3.4   Irrational numbers: The first crisis

The pentagram was the emblem of the Pythagoreans. We can only speculate about why they identified themselves with this fascinating five-pointed star, but their choice of logo was significant and had serious consequences.

The pentagram

The pentagram plays an important role both within mathematics and beyond. If we connect the outer points of a pentagram, a pentagon is formed, and similarly, a second pentagon connects the interior angles.

The number 5 appears surprisingly often in nature, and emerges most frequently in the form of a pentagram: A starfish has five arms and a starfruit has five lobes. If one cuts an apple in half crosswise, one observes that the core points in five directions.

The pentagram is not merely a form of expression in the natural world. It also plays a prominent role in art, or at least in representational art. Most stars on flags are pentagrams: Europe, Turkey, and the USA all display pentagrams on their flags. Hotel stars, some sheriffs' stars and Christmas stars are five-pointed.

We don't know why the Pythagoreans made the pentagram their emblem. We only know they did, and thus experienced a catastrophy.

To the Pythagoreans it was obvious to ask what the relationship is between the individual lengths in a pentagram. For instance, what is the ratio between the length of the line segment connecting the leftmost point of the pentagram to one of the inner vertices, and the line segment connecting this inner vertex to the rightmost point of the pentagram (see the following figure)?

The pentagram and the golden ratio

One might guess that these two lengths are related by the proportion 3 : 2. This is almost correct. The proportion 8 : 5 is more accurate, but is still not the exact value. The proportion 13 : 8 is better; this relationship differs by less than 1 percent from the true value.

In truth, you will never be able to express the relationship between these two lengths as a relationship between whole numbers. Never.

In other words, the algorithm for finding more and more accurate ratios will never end. One can always fit the shorter length into the longer length, and there will always be a remainder. This means that one cannot find a common measure for the two lengths: They are not commensurable; they are incommensurable.

We cannot verify this "negative" result by making measurements, but only by presenting a theoretically clean, crystal clear rationale. According to tradition, Hippasus of Metapontum, a student of Pythagoras, recognized this incommensurable ratio in the dodecahedron, a polyhedron with twelve pentagonal faces. Today we call this ratio an "irrational number"; that is, a number which is not a fraction. This finding shattered the Pythagoreans' belief that all ratios are rational. According to legend, Hippasus was "swallowed by the sea" for his insight.

Incidentally, we can calculate the sinister relationship between these two lengths in the pentagram—it is precisely $\frac{\sqrt{5}+1}{2}$, or approximately 1.618. Today we call this number "the golden ratio," which now has a thoroughly positive association.

The golden ratio was the first irrational number recognized by mankind. However, there are many more irrational numbers that are impossible to overlook, as Greek mathematicians soon realized. They discovered a characteristic which indicates that one is dealing with an irrational number: the radical sign (or roots, since at the time, the sign itself did not exist). Every root is irrational—aside from numbers such as $\sqrt{4}$. Specifically: The root of a natural number is either again a natural number or else it is irrational.

The proof that the number $\sqrt{2}$ is not a rational number, and thus cannot be written as a ratio of natural numbers, is famous. It is a proof by contradiction. This means: We assume that there exists a ratio of natural numbers equal to $\sqrt{2}$ and derive from this assumption a logical contradiction.

If a fraction equal to $\sqrt{2}$ exists at all, then there exists a fraction equal to $\sqrt{2}$ with a smallest denominator. (We consider some fraction equal to $\sqrt{2}$ and check which of the finitely many smaller denominators likewise result in a fraction equal to $\sqrt{2}$).

Let $\sqrt{2} = \frac{p}{q}$, where q is the smallest possible denominator of this ratio of natural numbers. The contradiction arises when we construct a ratio $\frac{m}{n}$ which is also equal to $\sqrt{2}$, yet whose denominator n is smaller than q.

Since $\sqrt{2}$ is the positive number whose square is equal to 2, we can square the above equation to get $2 = \frac{p^2}{q^2}$, and so

$$2q^2 = p^2.$$

We read this equation from left to right. Because the left side of the equation is an even number (namely 2 times a natural number), the right side, p^2, must also be an even number. Thus p must also be an even number. (If p were odd, then p times p would also be odd, according to the Pythagorean rule, "odd times odd is odd.") Therefore there exists a natural number m where $p = 2m$. We substitute this partial result into the previous equation and obtain $2q^2 = p^2 = (2m)^2 = 4m^2$.

If we divide this equation by 2, the result is

$$q^2 = 2m^2.$$

We read this equation from right to left: On the right is an even number, so the number on the left must also be even. Therefore q^2 is even and q must also be even. Consequently there exists a natural number n where $q = 2n$. Substituting this into the previous equation yields $(2n)^2 = q^2 = 2m^2$, therefore $4n^2 = 2m^2$ and thus $2n^2 = m^2$.

We solve this equation such that only the 2 remains on the left-hand side: $2 = \frac{m^2}{n^2} = \left(\frac{m}{n}\right)^2$. This means $\sqrt{2} = \frac{m}{n}$.

Now n is smaller than q because n is equal to $\frac{q}{2}$. Thus $\frac{m}{n}$ is a fraction that also represents $\sqrt{2}$, but whose denominator n is smaller than q. So, contrary to our original assumption, q was not the smallest possible denominator. This contradiction shows that our initial assumption was false. Thus there exists no ratio of natural numbers equal to $\sqrt{2}$.

This argument can easily be modified to show that the square root of every natural number that is not a perfect square must be irrational. This means that $\sqrt{2}, \sqrt{3}, \sqrt{5}, \sqrt{6}, \sqrt{7}, \sqrt{8}, \sqrt{10}, \sqrt{11}, \sqrt{12}, \ldots$ are all irrational numbers.

Similarly, the third root of any natural number that is not the third power of a natural number is irrational. Thus the third roots of $2, 3, 4, 5, 6, 7, 9, 10, 11, 12, 13, 14, \ldots$ are irrational. And so on with higher roots. This also yields infinitely many irrational numbers.

I cannot withhold from you another wonderful proof of the irrationality of $\sqrt{2}$. The proof was first found in the 1950s by the American mathematician Stanley Tennenbaum (1927–2005). Tennenbaum's proof begins in a similar manner to the classical proof: If there exist natural numbers p and q such that $\sqrt{2} = \frac{p}{q}$, then $p^2 = 2q^2$.

Tennenbaum now interprets this equation in a geometric way: There exists a square with sides of length p whose area p^2 is equal to the area $2q^2$ of two squares with sides of length q. Let p be the smallest natural number for which this is possible.

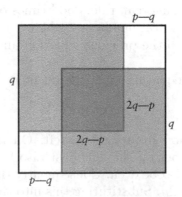

The irrationality of $\sqrt{2}$

Now we examine the partition of the square with sides of length p above. Because the total area $2q^2$ of the two small squares is equal to the area p^2 of the large square, the dark square in the center must be exactly as large as the two unshaded squares in the corners. We have thus found a smaller square, namely the dark square in the center, whose area is equal to the area of the two equal unshaded squares. Because we chose a minimal p, this is a contradiction.

3.5 Decimal numbers

At the beginning of the 16th century in Western Europe, a more or less peaceful conflict arose between Eastern and Western culture and science. The siege of Vienna by the Turks from September 27 to October 14, 1529, was a decisive time. In this context the news of "Turkish numbers" reached Europe. These "Turkish numbers" were none other than our decimal fractions.

The ultimate adoption of decimal fractions in Europe occurred thanks to the Flemish mathematician, physicist, and engineer Simon Stevin (1548–1620). In 1585 Stevin published his text *De Thiende* (The Tenth), in which he introduced the decimal fraction. In the text he wrote emphatically, "*Thiende* is the art of calculation wherein one executes all computations essential to mankind by means of whole numbers without fractions; it is found in the tenths sequence and consists of the digits by which a number is recorded."

Decimal numbers are natural numbers constructed of ones, tens, hundreds, and so on (see Chapter 2). A decimal fraction consists of a decimal number (which may be zero) "before the decimal point"; after the decimal point are tenths, hundredths, thousandths, and so on.

The number 0.4 thus signifies 4 tenths; the number 3.25 signifies 3 plus 2 tenths plus 5 hundredths, and 0.002 is simply 2 thousandths.

Stevin discovered that decimal fractions have enormous advantages over other fractions. In particular, one can calculate wonderfully with them because the rules for addition, subtraction, multiplication, and division of decimal numbers transfer easily to decimal fractions.

Many fractions can easily be converted into decimal fractions: $\frac{1}{2}$ is 0.5; $\frac{3}{10}$ is 0.3; $\frac{6}{25}$ is $\frac{24}{100}$, or 0.24. Other fractions such as $\frac{1}{3}$ or $\frac{5}{6}$ are difficult. Stevin was clear and radical on this point. His suggestion was to convert all measurements of weights, lengths, and times to the decimal system. There should be no subdivision into 6 or 12 smaller units, not to mention 60 minutes or 24 hours.

Stevin's prayers were not immediately answered—instead the conversion was delayed until the French Revolution. The French Revolution had a decisive impact on everyday life. The "revolutionary calendar" went into effect in 1792 and was in force until 1805. This calendar stipulated that the year should be divided into 12 months, each month consisting of 30 days; these 30 days were divided into 3 "decades," each composed of 10 days. This theoretically perfect system was brought into accord with reality by adding five "supplementary days" (and in leap years, six).

Even more radical was the division of a day into 10 hours, each hour containing 100 minutes, each minute composed of 100 seconds. Accordingly each day had 100,000 seconds, which only roughly agrees with the 86,400 "ordinary" seconds ($60 \cdot 60 \cdot 24$) in a day.

This apportionment never caught on, however, because all clocks would have been rendered unserviceable and everyone would have had to buy a new one.

In contrast, the 1000 millimeter meter, the 1000 milliliter liter, and the 1000 gram kilogram survived from the decimal revolution and are still in use today.

Decimal fractions have advantages compared to ordinary fractions—but they are not new numbers. We are still dealing with fractions! Every decimal fraction is a natural number plus a quantity of tenths plus a quantity of hundredths and so on. Tenths and hundredths are fractions, and the sum of two fractions is also a fraction!

The conversion of a decimal fraction into an ordinary fraction and vice versa is not always easy, but in principle these conversions are always possible. This task is performed according to the following rules:

- A finite decimal is easy to convert into a fraction: 0.12345 is equal to $\frac{12345}{100000}$.

- Conversely, if a fraction can be expanded into a fraction whose denominator is a power of ten, then one can read off the decimal. One expands $\frac{3}{200}$ into $\frac{15}{1000}$, and the decimal is 0.015.

- If a denominator cannot be expanded into a power of ten, then the corresponding decimal is not finite, but is instead infinite.

Indeed, the decimal ends in an infinitely repeated pattern called its *period*. An example is $\frac{1}{3} = 0.333\ldots$. We also write $0.\overline{3}$. Like all fractions, we can convert the fraction $\frac{1}{6}$ into a decimal fraction by performing the division $1 \div 6 = 0.1666\ldots = 0.1\overline{6}$.

- Conversely, we can reinterpret every periodic decimal fraction as an "ordinary" fraction. In general, this process is more difficult.

We can thus summarize: Every rational number can be represented both by a regular fraction and by a decimal, either as a finite decimal fraction or as a periodic decimal fraction.

Other decimal fractions exist that neither end nor become periodic. If, for example, one calculates $\sqrt{2}$, then one obtains $1.41421356237309504880\ldots$, and there is no periodic repetition in sight. Naturally one might imagine that this number ends much later in a periodic repetition of digits, however we know this does not happen. If the decimal representation of $\sqrt{2}$ ever became periodic then $\sqrt{2}$ would be a rational number, but we know that $\sqrt{2}$ is irrational.

These infinite and aperiodic decimal fractions lead us to a vast new empire of numbers, namely the real numbers.

Chapter 4

Transcendental numbers

4.1   The most mysterious number

When the Pythagoreans discovered irrational numbers in the 6th century B.C., they initially regarded such numbers as strange and unusual. However, they soon accepted the irrational numbers as an extension and enrichment of the known numbers and achieved familiarity with them. Numbers such as $\sqrt{2}$, $\sqrt{3}$, $\sqrt{7}$, or the golden ratio are indeed irrational and therefore difficult to calculate, but from a geometric perspective they are on an equal footing with the previously known numbers: For each of these irrational numbers we can construct a line of the corresponding length without much difficulty. It is thus clear that the irrational numbers truly exist. Even numbers that we cannot construct geometrically, such as $\sqrt[3]{2}$, can be imagined geometrically, namely—in the case of $\sqrt[3]{2}$—as the edge length of a cube of volume 2.

Over time we become used to expressions such as $\sqrt{5}$ or $\sqrt[7]{28}$. Thus we regard such symbols as "established numbers" and defer the calculation of their value. We can calculate wonderfully with such symbols without worrying about the value. For example, $\sqrt{2} \cdot \sqrt{3} = \sqrt{2 \cdot 3} = \sqrt{6}$. Such methods of calculation are seductive: Because we can calculate with roots as with the previously known numbers, we thus accept roots as numbers.

However one number seemed utterly simple yet eluded all attempts to describe it. This number seemed wholly natural, yet mathematicians could not master it, because no one could draw a line representing its length. The number was like a meteor found cool upon

61

the Earth which chronicles distant, unreachable worlds. This number, which mathematicians have studied for centuries, is the circle number π.

The number π shows that in addition to the "simple geometric" irrational numbers there are also numbers that cannot be described as roots or solutions to algebraic equations, but which go beyond—that is, "transcend"—ordinary irrational numbers.

The circle is one of the most elementary geometric shapes. It plays an extraordinarily important role not just in mathematics but also in daily life. Just think of the many circular objects in nature and technology. The most prominent representative is surely the wheel, but fruit and flowers, sun and moon, plates and cups, rings and crowns, bagels and burgers display the variety of circles. It is clear that calculations of the circumference and the area of a circle are among the most basic of mathematical tasks.

The circumference of a circle depends upon its diameter, and in a very simple way: The relationship of circumference to diameter is the same for every circle. If we measure the circumference and the diameter of a circle and then divide the circumference by the diameter, the ratio is always the same; thanks to Euler this number bears the name π ("pi"). Conversely we can calculate the circumference of a circle by multiplying the diameter by π.

Even in antiquity, mathematicians tried to precisely determine the number π. Around 2000 B.C. the Babylonians used $\pi = 3$ or $\pi = 3 + \frac{1}{8} = 3.125$ in their calculations. The Rhind papyrus (c. 1650–1600 B.C.) presents the approximation $\pi = \left(\frac{16}{9}\right)^2 = 3.1604\ldots$. The Bible also references π: In a description in 1 Kings 7:23 the reader learns that a circular ritual basin has a circumference of 30 cubits and a diameter of 10 cubits; therefore, biblical calculations were performed using $\pi = 3$.

The problem with all of these geometric methods is that we can imagine the circumference of a circle but we cannot—if we are honest—measure it precisely. We can only measure straight lines. Every curved line which we wish to measure precisely must be "rectified," that is transformed into a straight line of the same length.

The great mathematician and physicist Archimedes (287–212 B.C.) was probably the first to suspect that we will never determine π exactly. His method was ingenious, for he made the problem easy. Because Archimedes could not calculate the circumference of his circle, he drew a regular hexagon around the circle. This hexagon had a larger circumference than the circle—but he could calculate its circumference precisely!

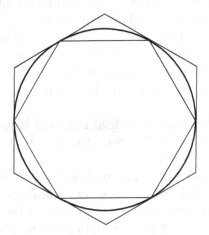

Circle with inscribed and circumscribed hexagons

Likewise, Archimedes drew a regular hexagon inside the circle; this hexagon had a smaller circumference than the circle. In this way, Archimedes obtained one upper and one lower bound for π. He didn't stop with the hexagon, but instead pursued his method to a regular polygon with 96 edges. This led him to the famous estimate:

$$3\frac{10}{71} < \pi < 3\frac{1}{7}.$$

If we convert this into decimal notation we obtain the first decimal places of $\pi = 3.14\ldots$.

You might ask, does π ever end? Does the decimal representation of π extend forever? The answer was given by the mathematician Heinrich Lambert, who proved in 1761 that π is an irrational number. This means that the decimal representation of π extends forever and never ends in a periodically repeating pattern. More than a century later, in 1882, Ferdinand Lindemann proved that π is transcendental, which means there is no algebraic equation with π as a solution. In particular, π is not a "radical," i.e., not an expression containing roots.

The search for additional decimal places of π is a theme that permeates the history of mathematics. Since computers can perform such calculations and very efficient series expansions for π have been discovered, the number of known decimal places for π has increased enormously. At present π is known to an incredible 10 trillion decimal places. Yet even this is nothing compared to infinitely many decimal places, and still each new digit is a surprise!

The first irrational numbers, the roots, were considered mysterious because it was difficult to calculate their precise values. However, one could realize them as lengths or describe them as radicals, and their existence was not doubted. Transcendental numbers are very different.

The definition of transcendental numbers is essentially a capitulation; it is a "negative definition," which only describes what transcendental numbers are not.

Apart from π, the irrational numbers we have learned about thus far are radical expressions. These are numbers such as $\sqrt{2}$, $\sqrt[3]{5}$, and $\sqrt{2} + \sqrt{3}$. Such numbers can be described by an algebraic equation. For example $x^2 = 2$ is an equation for which $\sqrt{2}$ is a solution. Analogously, $x^3 = 5$ and $x^4 - 10x^2 + 1 = 0$ are equations with respective solutions $\sqrt[3]{5}$ and $\sqrt{2} + \sqrt{3}$. Such numbers are called "algebraic numbers" because they are solutions of an algebraic equation. As "algebraic equations" we allow equations with an unknown x, whose coefficients (the numbers multiplying x, x^2, x^3 and so on) are rational numbers.

"Transcendental numbers" are those numbers that are not solutions of an algebraic equation with rational coefficients. We do not specify what transcendental numbers truly are, but instead merely state what they are not, namely solutions of an algebraic equation. Transcendental numbers go beyond the numbers previously imagined. In particular, π forces us to take transcendental numbers seriously.

At first mathematicians had no idea how to describe transcendental numbers in general or how many such numbers exist. Must we consider individual cases or is this a mass phenomenon?

These questions will be discussed in the following two sections.

4.2   Limits

If we wish to study the irrational and in particular the transcendental numbers properly, we must somehow get a grip on them. The perfect tool for this is the limit. The scrupulous handling of limits and therefore of the "infinitely small" is one of the greatest cultural achievements of humankind. Mathematicians struggled for approximately 2000 years to master limits, and thereby the real numbers, in a scientifically irreproachable manner.

Limits are necessary because irrational numbers in general can only be described as limits. Mathematicians have introduced symbols to represent some transcendental numbers such as π or e, but this notation is in essence mathematically meaningless.

In order to define a limit, first of all we need a sequence of numbers. Normally these are rational numbers, or fractions. For example, imagine the sequence $\frac{1}{1}, \frac{1}{2}, \frac{1}{3}, \frac{1}{4}, \frac{1}{5}, \frac{1}{6}, \ldots$. This sequence involves infinitely many numbers, and we know what these numbers are: For example the thousandth number of the sequence is $\frac{1}{1000}$. Mathematicians say that such a sequence "converges" if it approaches arbitrarily closely a number which we call the "limit." More precisely, imagine an arbitrarily small quantity. The question is then whether or not the numbers of the sequence ever differ from the limit by less than this infinitesimal quantity. This is assuredly not always the case! At the beginning of the sequence the numbers may behave as they wish; the beginning of the sequence has nothing to do with the limit. But at some point, after some particular term of the sequence, the numbers might differ from the limit by at most the chosen infinitesimal quantity. If this is true, then we say that the sequence "converges" to the limit.

The sequence $\frac{1}{1}, \frac{1}{2}, \frac{1}{3}, \frac{1}{4}, \frac{1}{5}, \frac{1}{6}, \ldots$ has as its limit the number 0. The numbers of the sequence not only approach 0 ever more closely (they also approach the number -1 ever more closely) but they also approach the number 0 more closely than any infinitesimally small distance: If we choose as our arbitrarily small distance the number $\frac{1}{1000}$, then all elements of the sequence beyond the thousandth element differ from 0 by less than $\frac{1}{1000}$ (because beyond the thousandth element the numbers in the sequence are $\frac{1}{1001}, \frac{1}{1002}, \frac{1}{1003}, \ldots$).

Furthermore, the elements of a sequence don't need to be positive: the sequence $-\frac{1}{2}, -\frac{1}{3}, -\frac{1}{4}, -\frac{1}{5}, -\frac{1}{6}, \ldots$ also has the limit 0.

Naturally, limits were not introduced in order to describe familiar numbers such as 0. On the contrary, the advantage of limits is that we can use them to describe numbers we cannot grasp directly, or numbers we can *only* define using limits. By constructing a limit we approach a number gradually; coming closer and closer, even arbitrarily close—but in general we will never actually reach it.

Limits are nothing new. Simple limits were considered by early mathematicians. The famous story about Achilles and the tortoise comes from Zeno (c. 490–430 B.C.). The story tells of a race in which two runners compete against one another in a knock-out system; the winner advances to the next round and the loser drops out of the tournament. Achilles represents the fastest runner, who is expected to win the tournament, while the tortoise represents the slowest runner, whom everyone else is expected to beat. Lots are drawn and Achilles and the tortoise must compete against one another. The tortoise knows: If he and Achilles start to run, this will be his last race. The tortoise thus tries to engage Achilles in conversation. In an innocent manner the tortoise asks, "Dear Achilles, you are surely willing to give me a bit of a head start?"—"As much as you like," is Achilles' generous answer. "100 cubits is enough," says the tortoise humbly. "I will catch up in no time," Achilles asserts from above. "Not in no time," replies the tortoise, "because while you run 100 cubits, I run too. I don't run as fast as you. But while you run 100 cubits, I move 10 cubits forwards."—"I can manage that instantly."—"No, because while you run ten cubits I move one cubit." Achilles is uncertain: "You mean, each time I reach the place where you were, you have moved forwards a little bit?" The tortoise waits, confident that his words will give Achilles pause: "Then I will never catch up with you?"

We don't know how the story ended. From a modern, mathematical perspective Zeno's paradox is easy to solve: Achilles and the tortoise are trying to determine when Achilles will overtake the tortoise; that will be at 111.111... cubits.

This is a limit. With finitely many iterative steps one cannot determine the limit exactly, but with real steps one can cross the limit. When Achilles has run 112 cubits he will have overtaken the tortoise.

Limits appear in Archimedes' work as well, and indeed even in mathematical formulation; for example, Archimedes knew that the infinite sum $1 + \frac{1}{2} + \frac{1}{4} + \frac{1}{8} + \dots$ is equal to 2.

As a scientist, the medieval theologian Nicolas d'Oresme (1330–1382) was interested in almost everything. He determined the limits of complicated infinite sums such as $\frac{1}{2} + \frac{2}{4} + \frac{3}{8} + \frac{4}{16} + \dots$. He represented this sum geometrically, rearranged it in a resourceful way, and saw that his series is the same as Archimedes' series $1 + \frac{1}{2} + \frac{1}{4} + \frac{1}{8} + \dots$—and thus surprisingly also has the value 2.

Oresme also solved another problem. There are infinite sums which one might assume should have a value, or converge, yet which do not in fact converge. The paradigm for these is the infinite sum $\frac{1}{2} + \frac{1}{3} + \frac{1}{4} + \frac{1}{5} + \cdots$.

Oresme proved that this series (or sum) becomes arbitrarily large and therefore does not converge to a fixed number. We can see this in the following way: Initially we examine the first element, $\frac{1}{2}$. The next two elements together yield at least $\frac{1}{2}$, because $\frac{1}{3} + \frac{1}{4} > \frac{1}{4} + \frac{1}{4} = \frac{1}{2}$. The next four elements together again yield at least $\frac{1}{2}$, because $\frac{1}{5} + \frac{1}{6} + \frac{1}{7} + \frac{1}{8} > \frac{1}{8} + \frac{1}{8} + \frac{1}{8} + \frac{1}{8} = \frac{1}{2}$. The following eight elements again give a total of at least $\frac{1}{2}$. And so on. The infinite sum $\frac{1}{2} + \frac{1}{3} + \frac{1}{4} + \frac{1}{5} + \cdots$ can be parceled out such that the sum of each part is larger than $\frac{1}{2}$, and because there are infinitely many such parts, the series cannot converge.

Through limits we obtain the real numbers. Every real number is the limit of a convergent sequence of rational numbers, and every convergent sequence of rational numbers has as its limit a real number. The real numbers consolidate the rational and irrational numbers, and indeed include both the "tame" algebraic and the "truculent" transcendental numbers.

We can describe some real numbers briefly, for instance through expressions such as $\sqrt{2}$, $\sqrt{13}$, $\sqrt[7]{359}$, $0.\overline{4}$, $7.\overline{351}$, "length of the diagonal of a regular pentagon of side length one," and so on. However, these are exceptions. Most real numbers have no "meaning" which we could use to characterize them briefly. Instead we need infinitely many symbols in order to define them, namely the infinitely many numbers of a convergent sequence whose limit is the number in question. This means that we can describe most real numbers only indirectly, through other numbers. In fact, most real numbers exist only as limits.

Let us take a step back. Limits of sequences of fractions can in turn be fractions (or whole numbers). For example, the sequence $\frac{1}{2}, \frac{2}{3}, \frac{3}{4}, \frac{4}{5}, \cdots$ converges to the number 1. However, these too are exceptions: The limits of almost all sequences of rational numbers are irrational. Yes, we can define all irrational numbers as limits of sequences of fractions. In short: The real numbers are exactly the limits of converging sequences of rational numbers.

An important real number which can only be obtained as a limit is Euler's number, e. Let us clarify this by considering the following thought experiment. Imagine that you have invested one dollar at the sensational interest rate of 100%. After one year you have two dollars.

However, you could invest more cleverly and withdraw the money after half a year; with interest you now have \$1.50. Now you invest this money in another bank—also with an interest rate of 100%. The interest rate is applied to the entire \$1.50 and at the end of the year you have \$1.50 + \$0.75 = \$2.25. Now you have tasted blood: You withdraw your money every three months and take it to a new bank. At the end of the year you now have \$2.44. If you withdraw your money once a month and seek out a new bank each time, by the end of the year your fortune will total \$2.61.

As a mathematician you would divide the year into n parts of equal length. At the end of each time period you would withdraw your money, including interest, and take it to another bank. At the end of the year you will have exactly $\left(1 + \frac{1}{n}\right)^n$ dollars. Now you can choose the number n to be arbitrarily large. Then you are interested in the limit of the sequence $\left(1 + \frac{1}{n}\right)^n$:

$$\left(1 + \frac{1}{1}\right)^1 = 2, \left(1 + \frac{1}{2}\right)^2 = 2.25, \left(1 + \frac{1}{3}\right)^3 = 2.44, \ldots .$$

This limit is $2.71828\ldots$. This means that no matter how energetically you move your money around you will never earn more than \$2.718... in a year.

This number, or the limit of the sequence $\left(1 + \frac{1}{n}\right)^n$, is called Euler's number and is abbreviated with e. This number provides the base of the "natural logarithm."

One might think of constructing sequences of real numbers and their limits with the goal of obtaining many more new numbers. However, this won't work. The real numbers are closed: Limits of sequences of real numbers are also real numbers. One cannot escape the real numbers by calculating the limits of sequences of real numbers. All of this sounds quite complicated. It *is* complicated: Many generations of mathematicians have grappled with these relationships.

If you think, "this is *too* complicated," then there is a solution for you: Simply imagine the real numbers as decimal fractions. Every decimal fraction—independent of whether it is finite, periodic, or infinite and aperiodic—is a real number. Conversely, every real number is a decimal fraction.

However, viewed correctly every decimal fraction is a limit. More precisely, every infinite decimal fraction is a convergent sequence of numbers. This sounds more difficult than it is: If we read a decimal fraction digit by digit we approach ever more closely the number that this decimal fraction represents. Most real numbers can only be described by writing or reading aloud their infinitely many digits.

Consider for example the number π. We know that $\pi = 3.14159\dots$. If we consider only the first digit after the decimal we obtain the number 3.1. This differs from π by less than one tenth (because π is larger than 3.1 and smaller than 3.2). If we regard the first two digits after the decimal, and consider 3.14, then this differs from π by less than one hundredth (because π is larger than 3.14 and smaller than 3.15). And so on. We obtain a sequence of numbers, namely

$$3, \ 3.1, \ 3.14, \ 3.141, \ 3.1415, \ 3.14159, \dots .$$

This sequence of numbers converges to π.

For every decimal fraction there is a real number. Sometimes two different decimal fractions represent the same real number. The typical example of this is the infinite periodic decimal fraction 0.999... which is also denoted $0.\bar{9}$. A mathematical theorem says that this decimal is equal to 1. Many people are admittedly of the opinion that this conclusion is false. They argue that although the number 0.999... does indeed approach 1 always more closely, it always remains a little smaller than 1.

To settle this issue we must initially realize that no number ever "approaches" another number. One number can only be smaller, equal to, or larger than another number. We get to the core of the matter when we consider what we mean by the symbol $0.\bar{9}$. This is the limit of the sequence $0.9, 0.99, 0.999, \dots$. Naturally each individual element of the sequence is smaller than 1, but the limit is exactly 1. We can therefore in good conscience write: $0.\bar{9} = 1$.

Again the question arises: What is a number?

If we say limits of convergent sequences are numbers, then the answer is: Numbers are the real numbers. The real numbers are those numbers which we need if we wish the limits of number sequences also to be numbers.

In addition, real numbers are indispensible for calculus and physics. If we want to describe movement mathematically then the rational numbers are inadequate, and we need the real numbers.

4.3   How many transcendental numbers are there?

We have seen that there are infinitely many irrational numbers, and at least a few transcendental numbers. Are there infinitely many transcendental numbers? What is the nature of a "typical" real number? Is it rational? Or algebraic? Or maybe transcendental? Is this question even meaningful?

Behind the question of meaning lurks another, fundamental question: Can we compare infinities according to size—or must we content ourselves with a banal "infinite is infinite"?

Enter the German mathematician Georg Cantor (1845–1918). He is often called the inventor of set theory, and many may suppose that Cantor spent his time playing with sets of objects. However, this would be the wrong idea. Cantor's subject was infinite sets and the resulting questions: What does it mean to say that two infinite sets contain equally many elements? What does it mean for one infinite set to be larger than another?

Cantor's thoughts were ignited by precisely the question of whether there are infinitely many transcendental numbers.

The key concept Cantor used to clarify this question is called "one-to-one correspondence," which refers to a one-to-one relationship between sets. We will make this clear with an everyday example: When a table has been properly laid each plate is associated with one fork, one knife, and one water glass. It is then clear that there are equal numbers of plates, forks, knives, and glasses on the table. This results directly from the unique correspondence between the items in each place setting on the table; one sees "at a glance" that there are equal numbers of each type of item, and does not need to count.

Cantor transferred this idea to infinite sets. The first degree of infinity, which according to many scientists is the only degree that one can imagine, is enumerable. Mathematicians get more specific. We say that a set is "countable" if we can arrange the elements of the set into a list such that there are a first, a second, and a third element, and so on, and such that all elements of the set appear in this arrangement—in short, if we can number the set using the natural numbers.

Every subset of the natural numbers is countable. For example, the set of all positive even numbers is countable. We can write them down in turn: $2, 4, 6, 8, 10, \ldots$ and it is clear that there is a first element, a second, etc., and that in this way all positive even numbers are included.

Proving the countability of the positive and negative whole numbers requires somewhat more finesse. We cannot order these numbers according to size, because there is no smallest number and one doesn't know where to begin.

The trick is to begin in the middle of the whole numbers, at the number 0, and then take one step to the right and one step to the left respectively. The arrangement is thus the following: $0, 1, -1, 2, -2, 3, -3, \ldots$ Once again it is obvious that every whole number occurs once in this list.

Cantor's first coup was the proof that the rational numbers are also countable. This is a markedly counterintuitive result, because naively one thinks there are "many more" fractions than natural numbers. After all, between each pair of natural numbers there are infinitely many fractions. Here we show that the fractions between 0 and 1 are countable; the rationale that all fractions are countable is merely a little more technically complicated. Cantor's idea was to sort the fractions according to the size of their denominator (see the figure below). First is the fraction with denominator 2, then the fractions with denominator 3, then those with denominator 4, and so on. The diagram now looks like the following:

$$1/2$$

$$1/3 \quad 2/3$$

$$1/4 \quad 2/4 \quad 3/4$$

$$1/5 \quad 2/5 \quad 3/5 \quad 4/5$$

$$1/6 \quad 2/6 \quad 3/6 \quad 4/6 \quad 5/6$$

The fractions that can be reduced (these are the fractions not in bold above) are ignored because the corresponding fractions already appear in one of the previous rows.

If one reads the array from top to bottom and each row from left to right, it is clear that the rational numbers are countable. Every rational number appears in some row. We can count the fractions row by row, thus constructing a one-to-one correspondence between them and the natural numbers. The rational numbers are thus countable.

With similar arguments Cantor showed in 1873 that the set of algebraic numbers is also countable. His real stroke of genius was however his second coup: In 1873 he also showed that the real numbers cannot be counted. They are uncountable, and this means that there are substantially more real numbers than algebraic numbers. In particular it follows that infinitely many transcendental numbers exist. In fact, essentially 100% of the real numbers are transcendental!

The argument for this is clear: Because the set of all real numbers is uncountable, but the subset of the algebraic numbers is merely countable, "the rest" (that is the set of transcendental numbers) must be uncountable and in particular infinite.

This realization was a sensation, because until then only special numbers had been proven to be transcendental. The most prominent examples are the Liouville number, whose transcendence Joseph Liouville (1809–1882) proved in 1844, and Euler's number e, whose transcendence was demonstrated by Charles Hermite (1822–1901) in 1873. Both proofs are difficult and utilize the special characteristics of these numbers. Cantor, in contrast, showed with a very simple method (almost sleight of hand) that there are infinitely many transcendental numbers.

How did Cantor prove the real numbers are uncountable? He had to demonstrate that one cannot count the real numbers. Cantor confined himself to the real numbers between zero and one, or all decimals that begin with "zero point something," and showed that even these are uncountable.

The proof is a sort of game, a game between someone who claims to have counted the real numbers, and ourselves.

Let's imagine that someone claims to possess a list of the real numbers between 0 and 1. This individual can use any means to create such a list—they may even be equipped with extraterrestrial intelligence. Using any method they produce a list and claim that this list includes *all* the real numbers between 0 and 1.

The first numbers in the list could be, for example, the following:

$$0.101010101010\ldots$$
$$0.123456789123\ldots$$
$$0.260971552873\ldots$$
$$0.999976290095\ldots$$
$$0.000004529847\ldots$$

$$\vdots$$

We demonstrate that at least one number fails to appear on their list.

There doesn't seem to be any pattern in this list, but a pattern is not necessary. Cantor found an ingenious trick which allows us to specify a number that is guaranteed not to appear on the list. This new number begins likewise with "zero point something." The digits after the decimal point of this new number are determined one after the other:

1. In the first position after the decimal point is a digit different from the digit in the first position of the first number on the list. The first number in this list is 0.10101...; the digit 1 occupies the first position after the decimal point. For the first digit of the new number we may choose any digit except 1. For example, we might decide that the first digit should be 7, and the new number would begin with 0.7.

2. In the second position after the decimal point of the new number is a digit different from the second digit after the decimal point of the second number in the list. The second number in the list is 0.12345.... In the second position after the decimal point is the digit 2; for the new number we choose any other digit, for example 4. The new number thus begins with 0.74.

3. In the third position after the decimal point of the new number is a digit different from the digit in the third position after the decimal point of the third number on the list. The third digit after the decimal of the third number on the list is 0, and we choose for example 1. The new number then begins with 0.741.

4. For the fourth digit after the decimal point we choose a digit
different from that in the fourth position of the fourth number
on the list. The digit in the fourth number on the list is 9; we
could choose 3. The new number thus begins with 0.7413.

The method is clear: Step by step we construct a number that
is different from every number on the list. The new number differs
from the fourth number on the list by at least the fourth digit and
it differs from the hundredth number on the list by at the least the
hundredth digit. In general the new number is different from the nth
number on the list because these two numbers have different digits
in at least the nth place after the decimal point.

We have thus shown that the real numbers between 0 and 1 are
uncountable!

If we try to imagine the number line with all the real numbers,
how does it look?

Simply put: The real number line is full. Every point is occupied
by a real number. There are no "free" points. If we had a pointer
with a truly pointlike tip, with which we could indicate precisely one
point, then this pointer would always hit a real number.

This sounds banal, but it is simply astounding. Although the
rational numbers are "dense," there are many "points" on the real
number line that are not occupied by a rational number. Indeed, most
points on the number line are not occupied by a rational number. If
one aims a pointer at the number line, then in almost every case one
encounters neither a rational nor an algebraic number, but instead a
transcendental number.

The real numbers cause quite a headache. Humans can imagine
single numbers independent of whether the number in question is 0, 1,
2, 1001, $\sqrt{7}$, or π. If we wish to imagine infinitely many numbers, all
of these numbers must follow a pattern. For example we can imagine
all fractions or all roots (or at least believe that we can). But alas,
we remain stuck in the countable numbers.

We can thus only truly imagine a vanishing minority of all real
numbers. This is the bad news. The good news is that through mathematical analysis we can obtain at least some inkling of the universe
of transcendental numbers. We thus perceive that there is something
"entirely different," something that vastly transcends the horizon of
our prior experience.

Chapter 5

Imaginary and complex

We have not yet considered one important aspect of numbers, namely numbers as solutions of equations. However, equations are a universal and exceptionally important subject in mathematics. Equations arise in a wholly natural way in applied problems: How large is a piece of land if one knows its length and width? How much do 7 eggs cost if 6 eggs cost $2.10? When will the human population on earth exceed ten billion? One can recast such questions as equations.

It is one thing to assemble equations, and quite another to solve them. The solving of equations runs through the history of mathematics like a guiding theme. The most important questions are:

- How can we solve an equation?

- How many solutions does an equation have?

- Does every equation have a solution?

Behind all these questions lurks the pivotal question: What kind of numbers do we need to solve equations of a particular type?

This question in particular has tested our understanding of numbers and given decisive impetus to its development over and over again. This issue has also driven mathematicians to the realization that the question "what actually is a number?" is anything but easy to answer. As applications become ever more complex, the equations that arise become ever more complicated. And for every new type of equation we need a new sort of number.

5.1   Linear equations

Mathematicians from the high civilizations of antiquity, for example in ancient Egypt, studied equations carefully. In 1858 the Scottish lawyer A. H. Rhind purchased an antique papyrus scroll in Luxor, Egypt. Today the scroll is found in the British Museum in London under the name "Rhind papyrus," and is the most extensive document about ancient Egyptian mathematics ever discovered. The Rhind papyrus was created in approximately 1650–1600 B.C., but is a transcript of a document 200 years older.

In the Rhind papyrus there are many problems that lead to equations. One means of solving such equations is the method of false position. Mathematically speaking the "false position" is an unknown quantity, or in modern terms the variable x.

Exercise 26 from the Rhind papyrus asks: A quantity and one quarter of that quantity together equal 15. What is the quantity?

The method of false position is simple, ingeniously simple. The idea of this method is to make life easy for ourselves. Imagine a solution: any hypothetical solution, simply a number with which we can calculate easily. For example, choose the number 4 as a hypothetical solution. Together, 4 plus 1/4 of itself equals 5.

But 5 is not the desired result. The result we seek is 15, not 5. The desired solution is therefore three times as large as our hypothetical solution. We thus multiply this hypothetical solution by three to obtain 3 times 4 is equal to 12, the desired solution. (Because 12 plus 1/4 of itself is 12 plus 3, or 15.)

Today we would write an equation with the unknown x and obtain $x + \frac{1}{4} \cdot x = 15$. We would mechanically multiply this equation by 4 and obtain $4x + x = 60$. We can rewrite this as $5x = 60$, and then the solution $x = 12$ emerges automatically. But how much more charming is the Egyptian method!

Equations of this type are called "linear equations." Linear equations are those in which x appears, but x^2, x^3, and $\sqrt{x}$ do not. Examples are $10x = 7$ and $5x + 3 = 11$. There are all kinds of methods for solving linear equations. With practice we discover first of all that every linear equation has precisely one solution. Secondly, the solution is a fraction (or a whole number). For example, the above equations have the solutions $x = \frac{7}{10} = 0.7$ and $x = \frac{8}{5} = 1.6$.

Linear equations are simple. We can easily solve them, if we have fractions at our disposal. In order to solve all linear equations, we need the rational numbers.

Because linear equations are so easy to solve, they can fool us into thinking that solving equations in general is child's play. This is completely untrue! Our account of the solution of equations has in store one of the most dramatic events in the history of mathematics.

5.2 Quadratic equations

Quadratic equations are those in which x^2 appears but x^3 and higher powers do not. Early mathematicians studied and solved such equations. The definition of Euclid's "golden ratio" leads to a quadratic equation, which Euclid solved in a purely geometric way.

In the middle ages quadratic equations were solved in systematic ways. The mathematician Muḥammad ibn Mūsā al-Khwārizmī (c. 780–850; see Section 2.3), who lived in Baghdad, played an important role in the development of the field. His book *al-ǧabr* (more precisely *al-kitāb al-mukhtaṣar fī ḥisāb al-ǧabr wa'l-muqābala, Calculation by Completion and Balancing*) is a compendium of rules and examples. Al-Khwārizmī summarized the contemporary knowledge of algebra and increased the understanding of linear and quadratic equations. His method was geometric and was equivalent to what we today call "completing the square."

One exercise in al-Khwārizmī's book states, "a square and 10 roots yield 39 units." In modern language this means: x^2 ("a square") plus $10x$ (10 "roots" or 10 times the solution) is 39; in short, $x^2 + 10x = 39$.

If the task were, "a square yields 36 units," then the problem would be easy to solve: We seek the length of the side of a square with an area of 36 units. The solution to this simplified problem is clearly 6.

How did al-Khwārizmī solve the equation $x^2 + 10x = 39$? First he reformulated the equation as a geometric problem:

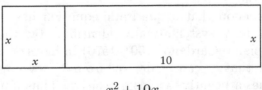

$$x^2 + 10x$$

Next al-Khwārizmī changed the figure in a small but decisive way: He divided the long rectangle with sides of length 10 and x into two new rectangles, each with sides of length 5 and x. He then moved the two new rectangles into position along two different sides of the small square with sides of length x. Finally, he completed the figure by adding a square with sides of length 5 to obtain a large square with sides of length $x + 5$.

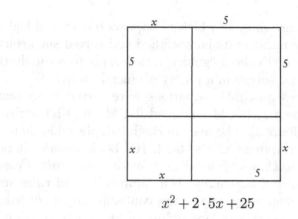

$$x^2 + 2 \cdot 5x + 25$$

The area of the added square is 25 units. The area of the original rectangle was $x^2 + 10x = 39$. The area of the new square is thus 25 units larger than 39, or $39 + 25 = 64$. The length of the side of the new square is thus 8, so $x + 5 = 8$, and so $x = 3$.

One problem with the geometric method is that naturally only positive solutions appear, because the solutions are lengths. In general, quadratic equations have two solutions, and one of these solutions is often negative. For example, the equation $x^2 - 3x - 10 = 0$ has one solution $x = 5$ and another $x = -2$. You can easily convince yourself that these solutions are correct by substituting the solutions into the equation.

The phenomenon that a quadratic equation has in general two solutions was first systematically identified by the renaissance polymath Gerolamo Cardano (1501–1576) in his *Ars Magna* (1545). Cardano wrote that "9 can be derived from 3 as well as −3, because a negative times a negative yields a positive. Thus if an even power is equal to a number, then its root has two values, one positive and one negative, which are equal to one another. If $x^2 = 9$, then $x = 3$ or $x = -3$."

Cardano accepted negative numbers as solutions, but only with a certain qualification, because he called a positive solution *vera* ("true") and called a negative solution *ficta* ("false").

The first to recognize negative numbers as numbers without constraint was the German theologian and mathematician Michael Stifel (1487–1567). He described a technique for solving quadratic equations, which essentially corresponds to our contemporary method using the quadratic formula. The contemporary method states:

The equation $ax^2 + bx + c = 0$ has the solutions

$$x = \frac{-b \pm \sqrt{b^2 - 4ac}}{2a}.$$

We can approach this monstrous formula in various ways. For example, we can first examine the most conspicuous features. The most distinctive feature is the root sign. The presence of the root sign means the solutions of a quadratic equation are very seldom whole or rational numbers; instead the solution usually requires a root. In order to solve quadratic equations, we must recognize irrational numbers, or more precisely square roots, as numbers.

The second conspicuous feature is the $\pm$. That means that we must either add the root expression to the first summand, $-b$, of the numerator or else subtract the same expression from the summand of the numerator. In general a quadratic equation has two different solutions. In mathematics the phrase "in general" means there may be exceptions. As it happens, if there is a negative number under the root sign, there is no solution at all (however, see Section 5.5). If the root is equal to zero then the solution is $-\frac{b}{2a} \pm 0$, or in short $-\frac{b}{2a}$, and in this case there is only one solution.

A second way to make the quadratic formula your own is to check whether it is correct. We do this by substituting numbers. Consider for example the equation $x^2 - 2x - 15 = 0$. For this equation the quadratic coefficient a is equal to 1, the linear coefficient b is equal to -2, and the constant term c is equal to -15. If we apply the quadratic formula, we obtain $\frac{2 \pm \sqrt{4 + 4 \cdot 15}}{2}$, which is 1 ± 4 and finally comes to 5 and -3. When we substitute these numbers into the original equation, we see that they are in fact solutions.

The French mathematician Françoise Viète (1540–1603), who called himself Franciscus Vieta in Latin, discovered a method by which the solution of a quadratic equation can often be easily determined. In order to solve al-Khwārizmī's equation $x^2 + 10x - 39 = 0$, one must simply find two numbers whose product is equal to -39 (the constant term) and whose sum is the negative of the coefficient of x, or the negative of 10. With a few attempts one sees that the solutions are equal to 3 and -13.

5.3   The drama of the cubic equation

In January 1535 something spectacular happened. A Venetian arithmetician named Niccolò Tartaglia (1500–1557) claimed that he could solve cubic equations, or equations containing the term x^3.

This was outrageous. Tartaglia was not the first; Scipione del Ferro (1465–1526) from Bologna claimed to have found a solution. However, del Ferro kept his solution strictly secret and shared it only with his student Antonio Maria Fior. Fior divulged the formula to no one—yet now Tartaglia also knew of a solution.

Tartaglia, of all people! Tartaglia led a life that was everything but ordinary. When he was 12 years old he was beaten by marauding soldiers until he looked "like a monster" and for a time could only stutter; hence the name Tartaglia, "the stutterer." Tartaglia later hid the injuries under a luxuriant beard. With tremendous diligence he taught himself mathematics, but he struggled to advance himself as an arithmetician in Venice.

This Tartaglia now claimed that he could solve cubic equations. The conflict was to be decided by a competition, a sort of mathematical duel between Tartaglia and Fior. In January 1535 the two mathematicians set one another 30 problems. In other words, each mathematician submitted cubic equations for which the other must find the solution. Tartaglia brooded over Fior's equations until, on the 12th of February, 1535, he found the canonical solution for one type of cubic equation and a day later a rule for solving a second type. In this way Tartaglia could solve all of Fior's problems within two hours.

Tartaglia had found a formula—it was a stroke of genius. However, this is not the end of the story. A short time later Gerolamo Cardano entered the scene. Cardano was one of the great mathematicians of his time. He knew everything, and what he didn't already know, he wanted to know. Cardano was currently working on his book *Ars Magna*, into which he intended to incorporate the collective knowledge of the times. Naturally, the formula for the solution of cubic equations must be included in this work.

Cardano sought Tartaglia out. Tartaglia had no chance against the great Cardano. Finally he agreed to a compromise: He revealed his methods to Cardano—under a pledge of secrecy, and only for scientific purposes.

Cardano seized his chance, and published Tartaglia's formula in *Ars Magna*. In some sense Cardano was fair: He wrote that Tartaglia provided the formula. However, the formula is known today as the Cardano formula. History is unjust.

In modern language we can describe the Cardano formula for solving a cubic equation in the following way: First of all we can recast every cubic equation such that it has the form $x^3 + px + q = 0$. This means that we can perform a substitution such that the summand containing x^2 disappears. Then for such an equation, the formula for the solution is:

$$\sqrt[3]{-\frac{q}{2} + \sqrt{\left(\frac{q}{2}\right)^2 + \left(\frac{p}{3}\right)^3}} + \sqrt[3]{-\frac{q}{2} - \sqrt{\left(\frac{q}{2}\right)^2 + \left(\frac{p}{3}\right)^3}}.$$

It is obvious that no one would offhandedly stumble across such a formula. In order to understand the formula at a rudimentary level, we first examine its external features.

First of all, the roots are the most conspicuous element. The large roots are third roots. However, this is not too bad, because the two third roots are almost the same; the second root expression differs from the first in that a plus sign is replaced by a minus sign.

With this formula, roots of negative numbers are inescapable. The individual roots can lead to "imaginary numbers"—even when the end result is a real number. Cardano calculated with these imaginary quantities as if they were numbers. It is a virtuosic calculation, but Cardano was flying blind. Instinctively he did everything right, but he didn't worry about the question of what these imaginary quantities actually are or whether they actually exist.

Today we use the letter i to indicate the square root of -1, or the number that, when squared, results in -1. We call numbers that include the imaginary unit i the "imaginary numbers." Today the term "complex numbers" prevails. For example, $3 + 5i$, $\frac{3}{5} - 8i$, and $4 + i\sqrt{2}$ are complex numbers.

We can also acquaint ourselves with the Cardano formula through an example: Consider the equation $x^3 + 6x + 2 = 0$. In this case $p = 6$

and $q = 2$. Now we must unravel the formula. The expression under the square root is:

$$\left(\frac{q}{2}\right)^2 + \left(\frac{p}{3}\right)^3 = 1^2 + 2^3 = 1 + 8 = 9.$$

The square root of 9 is 3.

The first large root is the third root of $-1 + 3 = 2$, and the second large root is the third root of $-1 - 3 = -4$. One solution of the equation is thus:

$$\sqrt[3]{2} + \sqrt[3]{-4} = \sqrt[3]{2} - \sqrt[3]{4} \approx -0.3275.$$

Now we can solve cubic equations. What next? This much is clear: After cubic equations come quartic equations, or equations containing terms of the form x^4. Quartic equations were solved by Lodovico Ferrari (1522–1565). The formula for the solution of quartic equations is a little more complicated than the formula for the solution of cubic equations, but this is only to be expected.

The next in line is the quintic equation ....

5.4   The tragedy of the quintic equation

The quintic equation—that is, an equation containing terms of x^5— offered resistance. Mathematicians struggled mightily to find a formula for the solution, but they ran into a wall. No one was successful.

Mathematicians were generous. The formula for the solution might well be complicated—it merely had to be a "radical expression." A radical expression is one that applies the four basic arithmetic operations and also takes roots of the coefficients of the equation. We may take square roots, third roots, or even thousandth roots, and use the basic arithmetic operations as often as we wish. We call this a radical expression. Mathematicians focused on radical expressions for two reasons. On the one hand, mathematicians were concerned with numbers they were able to envision easily. On the other hand, radical expressions led to formulas such as the quadratic formula and the Cardano formula.

The question was: Can every equation be solved by a radical expression? In particular, can quintic equations be solved by a radical expression?

It took a long time. It took a very long time, and the search ended sensationally. In the year 1799 the Italian mathematician and doctor Paolo Ruffini (1765–1822) published a horrifying result. He claimed that the general quintic equation was not solvable using roots. Unsolvable!

Ruffini's proof was flawed, but the spell was broken. In 1824 the Norwegian mathematician Niels Henrik Abel published a complete proof of the insolvability of the general quintic equation. Abel did not live long; he died of pulmonary tuberculosis in 1829 at 26 years of age—shortly before the offer of a post as a lecturer could reach him in the mail. Today the theorem that states the insolvability of the general quintic equation is rightly known as the "Abel-Ruffini Theorem."

This theorem was not only surprising, it was scandalous. It turned all previous conceptions upside down. Accepted opinion in the mathematical community was that every equation has a solution. The solution may be complicated and difficult to find, but that a solution exists was not seriously disputed. Now Abel had conclusively shown that the general quintic equation does not have a solution that can be represented as a radical expression.

The fundamental assumption that a radical expression has to provide a solution for each particular equation was certainly not yet off the table. For the proof that this fundamental assumption is flawed, we must thank another short-lived genius, the French mathematician Évariste Galois (1811–1832). Galois' life proceeded very dramatically. Frustrated by the outcome of the French Revolution, he became increasingly involved on the side of the Republicans. Yet his mathematical insight was far more revolutionary than his political affiliation. He developed a method whereby one can determine the structure of an equation and thereby its solvability. Galois recognized that one can assign every equation an algebraic structure (called a "group"), and that based on the structure of this "group" one can determine the solvability of the equation. Galois ventured into entirely new territory and had a premonition of visionary portents of modern mathematics far in advance of its development. He wrote a great treatise on this subject, which was not accepted for publication.

Galois was challenged to fight a duel on the morning of the 30th of May, 1832. The reason for the duel was his love affair with a young Frenchwoman who was already engaged to another. It remains unclear today whether this was a true conflict or whether the duel was staged by his political opponents. In any case, the night before

the duel Galois sat at his desk and wrote as if possessed a letter
to his friend Auguste Chevalier. Galois knew that time was running
out, for he sensed that he would not survive the duel. He knew also
that he had something to say to the world, even much to say. He
wrote frantically and tried to put to paper all of his mathematical thoughts. Galois' letter is a staggering mathematical document.
Again and again he scribbled in the margin of the letter, "I am out
of time."

Galois died on the 31st of May, 1832 from injuries sustained during
the duel. His vision lived on, though it would be decades before
mathematicians fully understood his genius.

5.5   Every equation can be solved!

Does every equation have a solution? The previous section might suggest that the answer is "no." Abel and Galois sought solutions that
are *radical expressions*, and thus constrained themselves to very special numbers, in particular to algebraic numbers. What if we dropped
this constraint and allowed solutions with *arbitrary* real numbers?
Does every equation then have a solution?

In order to answer this question we must clarify which equations
we have in mind and what type of numbers we will accept as solutions.

- Does every equation have a *natural* number as a solution? No,
  because the equation $2x + 4 = 0$ has only the negative number
  -2 as a solution.

- Does every equation have a *whole* number as a solution? No,
  because the equation $4x - 3 = 0$ has only the fraction $\frac{3}{4}$ as a
  solution.

- Does every equation have a *rational* number as a solution?
  No, because the equation $x^2 - 2 = 0$ has only the irrational
  numbers $\sqrt{2}$ and $-\sqrt{2}$ as solutions.

- Does every equation have a *real* number (a decimal) as a solution? No, because the equation $x^2 + 1 = 0$ has no solution that
  is a real number. This equation does, however, have a solution that is a complex number. If in place of x we substitute the complex number i (where $i = \sqrt{-1}$), the equation

works out neatly. Incidentally, $-i$ is also a solution, because $(-i)^2 = (-1)^2 \cdot i^2 = i^2 = -1$. That means: The equation $x^2 + 1 = 0$ has exactly two solutions, both of which are complex numbers.

- One last question: Does every equation have a *complex* number as a solution? The answer to this last question is: Yes! Every equation can be solved, if one allows complex numbers as solutions. This is the statement known as the "Fundamental Theorem of Algebra."

What is an "equation"? We consider equations with one unknown, which we usually call x. Examples of equations are:

$$x^2 - 1 = 0,$$

$$x^4 - 4x + 3 = 0,$$

$$x^5 - x + 1 = 0, \text{ or}$$

$$x^6 + \frac{1}{3} \cdot x^5 - \frac{7}{8} \cdot x^4 + 3x^2 + \pi = 0.$$

We list a few simple observations:

- In our examples, the right side of the equation is always zero. This serves simply to standardize the equations. One can always bring terms to the left side of an equation. Before we allowed negative numbers as solutions we had to arrange the equations such that all occurring numbers were positive. From the modern perspective, this led to a glut of unnecessary cases.

- In each instance there is a highest power of x. This power is called the *degree* of the equation: The equations above are of degree 2, 4, 5, and 6. Equations of the first degree are the linear equations that we discussed in Section 5.1; quadratic equations are equations of the second degree, cubics the third, quartics the fourth, and quintics the fifth degree.

The question of the solution or solutions of such an equation is a central theme in mathematics. The degree of an equation is therefore a central concept which we shall use to formulate the following results.

The first vision of the general solvability of equations originates with the Nuremberg arithmetician Peter Roth, who surmised in 1608 that an equation of the nth degree has at most n solutions. This conjecture was based on experience with equations of first, second, and third degree. However, it was René Descartes who proved Roth's result in 1637. His primary resource was the following insight:

Suppose an equation has a solution, in other words a number t for which the equation is valid. Then one can "split off" the factor $x - t$ from the equation. That means the following: Let us abbreviate the entire left side of the equation with f. Then Descartes states that if t is a solution, one can write f as $f = g \cdot (x - t)$. Thus $g = 0$ is an equation of one degree less, or an equation of degree $n - 1$.

For example the equation $x^3 - 2x^2 + 1 = 0$ has the solution $x = 1$, which can easily be verified by substitution. Then we split off the factor $x - 1$, and we see that $x^3 - 2x^2 + 1 = (x^2 - x - 1) \cdot (x - 1)$ is valid.

Descartes' theorem is very useful for solving equations: If we have found n solutions, we know we are finished. However, Descartes' proof of this theorem also provides enormously important insight from a theoretical as well as a practical viewpoint. If we have found one solution, we can reduce the equation $f = 0$ of degree n to an equation $g = 0$ of degree $n - 1$.

Like Descartes, we can use this method to prove the assertion that an equation of degree n has at most n solutions. We demonstrate this using a third degree equation. If this equation has no solution, then the theorem is valid. If the equation has one solution, we can split off $x - t$, and an equation of the second degree remains. This second degree equation has at most two solutions. The third degree equation thus has at most three solutions.

In the year 1629 the French mathematician Albert Girard went one substantial step further. He dared to assert that an equation of the nth degree has exactly n solutions. In particular this statement indicates that every equation has at least one solution. Girard had no evidence for his assertion. But he knew that his assertion could be true only if one also accepts "imaginary quantities" as solutions.

We discover this with quadratic equations. If, for example, we solve the equation $x^2 + 2x + 3$ using the quadratic formula, we obtain the solutions $-1 \pm \sqrt{-2}$. This example is not an exception, for the phenomenon occurs markedly often: Whenever the constant term c in the quadratic equation is larger than $\frac{b^2}{4a}$ then we obtain as solutions not real but complex numbers.

Complex numbers seemed and still seem very strange. In order to accept these numbers it is helpful to interpret them in a geometric

manner. We cannot interpret complex numbers geometrically with
the aid of the real number line alone, because the number line provides no place for imaginary quantities. However a geometric inter-
pretation works wonderfully well with a number *plane*.

A complex number has the form $a + bi$, where a and b are real
numbers. For example, $2 + 3i$ (where $a = 2$ and $b = 3$) and $5 - i$
(where $a = 5$ and $b = -1$) are complex numbers. We call a the real
part and b the imaginary part of the complex number $a + bi$.

We view a as the $x-$coordinate and b as the $y-$coordinate, and
the complex number is thus a point on the "complex plane."

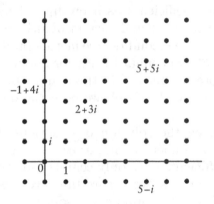

The complex plane

The first mathematician to interpret complex numbers as points
on a complex plane was the Norwegian and Danish mathematician
Caspar Wessel (1745–1818). This method was first generally recognized when Carl Friedrich Gauss (1777–1855), the "Prince of Mathematics," used this representation around 1811.

Girard's main argument for the validity of his theorem was aes-
thetic in nature. He believed that a "general rule" which stated that
every equation of nth degree has exactly n solutions would provide
certainty. This was naturally not a proof. His idea remained a conjecture, gaining no truth through an argument essentially external
to mathematics.

The Fundamental Theorem of Algebra was in truth formulated
by many mathematicians in different forms, but a formulation is not
a proof. The first consistent proof was constructed by the 22-year-
old Carl Friedrich Gauss in his dissertation in 1799, which allowed

him to graduate in absentia from the University of Helmstedt with a doctorate.

In his life Gauss constructed four proofs of the Fundamental Theorem of Algebra, the last on the occasion of the fiftieth anniversary of his doctoral degree in 1849. One proof would have been enough. Yet through his proofs Gauss not only demonstrated the importance of the Fundamental Theorem of Algebra, but also showed that this theorem is inextricably entwined with all fields of mathematics.

If you don't like imaginary quantities, you can formulate the theorem in a different way: One can write the left side of every equation with real coefficients as a product of factors of degree 1 (factors of the form $x - t$) and factors of degree 2 (factors of the form $ax^2 + bx + c$). The second degree factors produce complex solutions where applicable. Leonhard Euler formulated the Fundamental Theorem of Algebra in this way in the year 1748.

For the last time in this book we pose the question: What is a number?

When considering the solution of equations it makes sense to regard the complex numbers as the correct set of numbers. Equations with complex coefficients have only complex solutions. By introducing the complex numbers, we ensure that every equation with real or complex coefficients has numbers as solutions.

Additional Notes

For this English edition, the translators and the author have added a few observations that expand on the original text.

Section 1.1

Even today there are many cultures that do quite well without the ability to count. After all, why does one need to count the number of one's children? An excellent book by Alex Bellos, *Here's Looking at Euclid*, deals with such issues in its first chapter. In particular he talks about the Munduruku living in an area of the Amazon basin roughly the size of New Jersey, who have no words for numbers beyond five. For them, the idea of counting children—or in fact anything—is simply ludicrous.

Even more interesting is the way in which the Munduruku view the number line. American adults think of the number line as it is presented in this text where adjacent numbers are always the same distance apart, but the Munduruku when tested with various numbers of dots think that the intervals between numbers decrease as the numbers increase. This would seem but an anomaly if it weren't the case that when American kindergarten children are given the same test they also think that larger numbers are closer together than smaller numbers.

Section 1.2

In this section we see the value of thinking of numbers visually and use geometric arrangements of numbers to prove three interesting and useful facts about numbers: 1) the sum of any two odd numbers is an even number; 2) the sum of the first n odd numbers is the nth square number (for example, $1 + 3 + 5 + 7 + 9 = 5^2$); 3) the sum of the first n natural numbers is the nth triangular number (for example, $1 + 2 + 3 + 4 + 5 = 15$).

You might want to use similar "proofs without words" to prove the following facts:

a) the sum of any two consecutive triangular numbers is a square number (for example, $10 + 15 = 25$); hint: arrange a triangle representing 10 together with a triangle representing 15 in such a way that the resulting figure looks like a 5×5 square, then notice that you can do the same thing with *any* two consecutive triangular numbers.

b) a formula for the nth triangular number is given by $\frac{n(n+1)}{2}$ (for example, $\frac{5(6)}{2} = 15$); hint: put two triangles each representing 15 together in such a way that the resulting figure looks like a 5×6 rectangle, then notice that the same trick works for any value of n.

Section 1.3

There is a famous legend that Emperor Yu (around 2200 B.C.) was walking along the Luo river and saw a turtle emerge from the water with the pattern of this magic square on its shell! Luoshu literally translates as Luo River Writing. This legend as well as much else about magic squares can be found in Clifford Pickover's book *The Zen of Magic Squares, Circles, and Stars*, including a truly remarkable 16×16 magic square produced by Benjamin Franklin.

Section 1.4

As mentioned in the text, the largest known prime is $2^{57885161} - 1$. This is an example of what is called a *Mersenne* prime, that is, it is of the form $2^p - 1$ where p is also a prime. The search for ever larger primes is almost exclusively restricted to testing Mersenne primes and you can track this ongoing search (as well as discover why these are called Mersenne primes) at www.utm.edu/research/primes/mersenne.shtml. By the time you are reading this, an even larger prime number may have been found.

The Fundamental Theorem of Arithmetic not only guarantees that any natural number $n > 1$ is either itself prime or is a product of primes (this is almost obvious, since if n is not a prime it can be factored into a product of two smaller numbers a and b so that $n = a \cdot b$, but then if a and b are not themselves prime they can be similarly factored, and so on, until n is represented as a product of primes). The Fundamental Theorem of Arithmetic also guarantees that this factorization of n into a product of primes is unique. So, for example, the factorizations of the numbers 24, 999, 1000, and 1001 presented in the text are the only possible factorizations of these numbers into primes. The "uniqueness" of prime factorization is not at all difficult to prove and can be found in almost any book on number theory.

Euclid's proof that there are infinitely many primes is considered to be one of the most elegant proofs in all of mathematics. In the last 2000 years, mathematicians have found many beautiful proofs of Euclid's theorem. (For instance, six proofs appear in *Proofs from the Book* by Martin Aigner and Günter M. Ziegler.) One such proof was found as recently as 2006 and is every bit as elegant as Euclid's proof (see F. Saidak, A new proof of Euclid's theorem, *American Mathematical Monthly* 113(10) (December 2006), 937–38). This proof is based on the simple idea (also used in Euclid's proof) that no prime can divide both a number n and $n + 1$. So the new proof goes like this, begin with any natural number $n > 1$, then the number $n(n + 1)$ has at least two prime divisors. Similarly, then, the number $n(n + 1)(n(n + 1) + 1)$ has at least three prime divisors, and so on.

Section 1.5

Wiles's proof of Fermat's Last Theorem owes much to many other mathematicians. The final proof actually began in the 1950s with a conjecture connecting topology with number theory called the *Taniyama-Shimura Conjecture*. In 1984, Gerhard Frey showed that this conjecture was related to Fermat's Last Theorem. Then, in 1986, Ken Ribet proved that this conjecture implies Fermat's Last Theorem. Thus, Wiles produced a "proof" in 1993, which as is turns out had a serious flaw, but with the help of a former student, Richard Taylor, he was able to complete his proof based on a special case of the Taniyama-Shimura Conjecture.

Section 1.7

While obviously an impressive feat, the success of factoring the 232 digit number known as RSA-768 into a product of two primes is somewhat misleading since the task took two years and used many hundreds of computers working together. In 1991 RSA Laboratories produced a list of 54 numbers ranging in size from 100 to 617 digits, each having exactly two prime factors, and issued an open challenge to factor any of these numbers. To date, in addition to RSA-768, only the smallest 17 of these numbers have been factored. So, at least for now, an RSA encryption system based upon using two primes p and q with over 150 digits each appears to be safe.

Section 2.1

The Ishango bone is surely one of the earliest artifacts that suggest mathematical activity among humans. Not only have the prime numbers 11, 13, 17, and 19 been notched along one side of this bone; the other two sides contain numbers of notches that also appear to have some degree of mathematical significance. The markings on the Ishango bone have been widely interpreted through the years (see, for example, *The History of Mathematics: A Reader* edited by John Fauvel and Jeremy Gray).

Section 2.2

In an article "Arithmetic with Roman Numerals" (*The American Mathematical Monthly* 88(1) (January, 1981), pp. 29–32) James G. Kennedy provides various simple methods for doing basic arithmetic with Roman numerals. As Kennedy points out, the exact methods by which the Romans calculated have been lost, and his methods may well just be a modern invention.

Section 2.3

We now take it for granted that 0 is a number and fail to recognize what a historic leap this was in India almost three thousand years ago. After all, numbers are for counting and when was the last time you felt the need to have a number to represent how many pink elephants were in your bedroom? It is impossible to overstate the importance of this insight and the resulting Indian decimal system that rapidly passed through the Islamic world and eventually spread through Europe in the middle ages. Carl Friedrich Gauss, one of the great mathematicians of all time, could not understand how another of the great mathematicians Archimedes could have failed to invent the decimal system and regarded this oversight as one of the great calamities in the history of science, saying "to what heights would science have now been raised if Archimedes had made that discovery!"

Section 2.4

Of course, the most familiar case of the nines rule mentioned in this section is that a number is divisible by 9 if and only if its checksum (that is, the sum of its digits) is divisible by 9. This is a special case of the generalized nines rule that the remainder when dividing a number by 9 is equal to the remainder when dividing the checksum by 9.

There are also very handy divisibility rules besides the ones given in the text for 2 (the number 54 is even because its last digit 4 is even) and for 9 (the number 123,456,789 is divisible by 9 because the sum of its digits is divisible by 9). The rule for divisibility by 3 is almost the same, so for example 111,111 is divisible by 3 because the sum of its digits is divisible by 3, whereas, 1,111,111 is not divisible by 3. To test a number for divisibility by 11, you can test instead the alternating sum of its digits, so for example, 1,358,024,679 is divisible by 11 because $1 - 3 + 5 - 8 + 0 - 2 + 4 - 6 + 7 - 9 = -11$ is divisible by 11.

The reason that the *nines rule* works is that if we write an integer N as

$$d_1 + d_2 \cdot 10 + d_3 \cdot 10^2 + \cdots + d_{r+1} \cdot 10^r$$

in terms of its digits $d_1, d_2, \ldots, d_{r+1}$, then the sum of its digits is

$$S = d_1 + d_2 + \cdots + d_{r+1}.$$

So, the difference between the number N and the sum S of its digits is

$$N - S = d_2 \cdot 9 + d_3 \cdot 99 + \cdots + d_{r+1} \cdot 999 \ldots 9,$$

which is divisible by 9. Hence, N is divisible by 9 if and only if the sum S is divisible by 9.

Section 2.5

As humans we often view the world in binary terms: good/bad, dead/alive, right/wrong, hot/cold, left/right, black/white, conservative/liberal, winners/losers, and on and on, and yet we think our decimal number system is an almost perfect middle ground between (say) the Babylonian base 60 system that in effect requires 60 different symbols for each "decimal" place and the seemingly inefficient binary system that requires only two symbols but takes a huge number of "digits" just to represent a relatively small number such as one thousand as 1111110010 (not to mention that the square root of 2 now looks likes $\sqrt{2} = 1.01101010000010011\ldots$). Yet, now that computers control so much of our lives and computers store and transmit data by sequences of on and offs it is the binary system that is clearly now winning the day.

Section 2.6

The barcode described in the text as the European Article Number has now been largely replaced by what is known as the Universal Product Code. For this 12-digit barcode you add the odd digits, multiply by 3, and then add the even digits from 2 to 10, then check that the last digit of this sum agrees with the 12th digit, the check digit. You might want to try this on this next item you purchase.

Section 3.1

Babylonian mathematics was quite sophisticated. As early as around 1800 BC, they had the following sexagesimal approximation for $\sqrt{2}$: $1 + \frac{24}{60} + \frac{51}{60^2} + \frac{10}{60^3}$, which is accurate to five decimal places since $\sqrt{2} = 1.41421\ldots$. They used this approximation, for example, to find the diagonal of a 30×30 square. This, as well as considerable other evidence, shows that they knew the Pythagorean Theorem well before the time of Pythagoras.

Here is an example of a problem posed on a Babylonian tablet (again from around 1800 BC) that shows they could deal with fractions such as $\frac{1}{7}$, $\frac{1}{11}$, and $\frac{1}{13}$:

> I found a stone, but did not weigh it; after I subtracted one-seventh, added one-eleventh, and subtracted one-thirteenth, I weighed it as 1 ma-na. What was the original weight of the stone?

Why would the Egyptians use such a peculiar system of unit fractions? A possible answer can be found in a book by David Reimer, *Count like an Egyptian*. Almost one third of the Rhind papyrus (dating from around 1650 B.C.) is devoted to a list of the doubles of all unit fractions from $\frac{1}{3}$ to $\frac{1}{101}$. So, for example, $\frac{2}{7}$ is given as the sum of $\frac{1}{4}$ and $\frac{1}{28}$. This looks strange to us. But, an Egyptian scribe would have noticed that $\frac{1}{4}$ is a pretty good approximation to $\frac{2}{7}$. If we multiply $\frac{2}{7} = 2 \times \frac{1}{7}$ by 28 we get 2×4 which equals $7 + 1$, and dividing by 28 yields $\frac{1}{4} + \frac{1}{28}$, just as the Rhind papyrus tells us.

Section 3.4

Mathematics is perhaps unique among the scientific disciplines in that once a mathematical truth is discovered it remains true forever. Newton is famous for developing a remarkably useful theory of gravity, but his theory has now been replaced by a new theory of gravity described by Einstein without which, for example, the GPS location system on our iPhones would not work. In this section it is proved that $\sqrt{2}$ is irrational and in Section 1.4 Euclid's proof of the infinitude of primes was presented. The great British mathematician G. H. Hardy captured both the beauty and the endurable quality of both of these mathematical results which he described as "theorems of the highest class" and wrote that "each is as fresh and significant as when it was discovered—two thousand years have not written a wrinkle on either of them." And yet this section concludes with a lovely visual proof of the irrationality of $\sqrt{2}$ reminding us that even eternal truths can be looked at in new ways.

Section 3.5

It has been known for over two thousand years that not all numbers are rational. The most famous example is that the diagonal of a unit square, $\sqrt{2}$, is not a rational number. In this section we see a powerful argument as to why not all numbers are rational, simply because a rational number must have a decimal expansion that is either finite or periodic. Moreover, since it is easy to imagine that a random infinite decimal would be highly unlikely to be periodic, it is certainly plausible that the number of irrational numbers far exceeds the number of rational numbers. In other words, the rational numbers are but just a tiny proportion of all real numbers.

Section 4.1

Archimedes' brilliant idea of approximating the value of π using inscribed and circumscribed polygons with 96 sides led him to conclude that $3\frac{10}{71} < \pi < 3\frac{1}{7}$, and so we know the value of π to two decimal places: $\pi = 3.14$. In the fifth century the Chinese mathematician Zu Chongzhi used polygons with 24,576 sides to approximate π to seven decimal places. This was the best approximation ever found for π for the next nine hundred years.

Section 4.3

We have seen in this section that there are at least two sets having different sizes, the set of natural numbers and the set of real numbers. But Cantor didn't stop there. He showed that in fact there are *infinitely* many sets having different sizes. To do this he began with an obvious idea: for any finite set S, the set of all subsets of S is a bigger set than S. For example, if $S = \{a, b, c\}$ then S has 3 elements, whereas the set of all subsets of S is the set $\{\emptyset, \{a\}, \{b\}, \{c\}, \{a, b\}, \{a, c\}, \{b, c\}, \{a, b, c\}\}$ which has 8 elements (by the way, $\emptyset$ is the standard notation used for the "empty" set, that is, the set that has no elements). Cantor simply—but brilliantly—applied this same idea to infinite sets and showed that even if S is an infinite set then the set of all subsets of S is a bigger infinite set than S. He proved this using exactly the same argument used in this section to prove that the set of real numbers is bigger than the set of natural numbers. This clever argument is now—for fairly obvious reasons—called "Cantor's diagonal argument" (Cantor himself first used this argument in 1891 to show that the set of all infinite sequences of 0s and 1s is uncountable.) Cantor is now off and running: he begins with an infinite set S and builds a bigger infinite set T, which he then uses to build a still bigger infinite set, and on and on, building an infinity of infinite sets, each infinite set bigger than the previous one. So, one consequence of Cantor's brilliant idea is that there is no "biggest" infinite set.

There is one huge question that remains unresolved about these infinity of infinities. If you begin with the infinite set of natural numbers, then we know that the set of all subsets of the natural numbers is a bigger infinite set (in fact, surprisingly, it turns out that this new set is the same size as the set of real numbers). Might there be a set that is bigger than the set of natural numbers but smaller than the set of real numbers? This would seem likely, but no such set has ever been found! There is thus a famous conjecture, called "The Continuum Hypothesis" that says *no* such set exists.

An excellent book by Edward Burger and Michael Starbird, *The Heart of Mathematics*, has a wonderful chapter where they explore the general question "What does infinity mean?" in an extremely clear and imaginative way.

Section 5.2

It is worth noting that that Viète's method to solve quadratic equations such as $x^2 + 10x - 39 = 0$ is equivalent to our modern method of factoring to write $x^2 + 10x - 39 = (x - 3)(x + 13) = 0$, which in turn means that $x - 3 = 0$ or $x + 13 = 0$, and so $x = 3$ and $x = -13$ are the solutions.

Section 5.3

In the text it is mentioned that any cubic equation $ax^3 + bx^2 + cx + d = 0$ can be recast in the form $x^3 + px + q = 0$ and then gives a formula for the solution in terms of p and q. This step requires a clever trick but is very easy to do. First of all, you can divide the original equation by a, which means any cubic equation can be written in the form of $x^3 + bx^2 + cx + d = 0$. Then we use a substitution $x = y - \frac{b}{3}$, which eliminates the y^2 term (you should do this for yourself), and so we are left with a cubic equation of the form $y^3 + py + q = 0$. If you can solve this latter equation in terms of y you automatically have solved the original equation in terms of x.

Section 5.5

There is a nice geometric way of looking at the fact that an equation of degree n can have at most n solutions. For example, a cubic equation equation such as $x^3 - 2x^2 - x + 2 = 0$ can have at most three solutions because when we graph the cubic polynomial $x^3 - 2x^2 - x + 2$ it can cross the x-axis at most three times; in fact, a cubic polynomial results in three solutions if it crosses the axis three times, or one solution if it crosses the axis only once, or two solutions if the axis is crossed once and is tangent to the curve at another point.

Girard's assertion that an equation of the nth degree has exactly n solutions is correct only if we count some solutions more than once. For example, the equation $x^3 - 5x^2 + 7x - 3 = 0$ has two solutions, $x = 1$ and $x = 3$, but if we write $x^3 - 5x^2 + 7x - 3 = (x - 1)(x - 1)(x - 3) = 0$ we see that the solution $x = 1$ occurs twice, and in this sense Girard's assertion is correct in that we have three solutions: $x = 1$, $x = 1$, $x = 3$.

Addendum

This book ends in a very satisfying manner, beginning with just the natural numbers, the number line has through the years gradually been filled in to include negative numbers, rational numbers, irrational and even transcendental numbers. So we are left with the impression that the real line we draw (as an x-axis, say) is indeed an utterly solid continuum of points each of which represents one of these of numbers.

Nonetheless, mathematicians have invented other number systems. Yet the underlying idea in each of these systems has always adhered to Dedekind's view that a number system is a set of objects one can calculate with (such as by adding or multiplying). For example, in 1843, an important shift in what we might call a number came when the Irish mathematician William Rowan Hamilton discovered the *quaternions*, a number system that extends complex numbers. Just as complex numbers represent two-dimensional space, the quaternions represent four-dimensional space. A quaternion is a number of the form $a + bi + cj + dk$ where a, b, c, and d are real numbers and i, j, and k satisfy the formulas $i^2 = j^2 = k^2 = ijk = -1$. However, the quaternions contain a very interesting surprise for us: they are not commutative under multiplication, for example $ij \neq ji$ (by the above formulas, $ij = -ijk^2 = -(-k) = k$, but $ji = -jiijk = j^2 k = -k$).

A number system called the *p-adic numbers* (for any prime p) was discovered in 1897 by the German mathematician Kurt Hensel and is especially interesting because the *p*-adic numbers contains the rational numbers but are not contained in either the real or complex numbers. In other words, they extend the rationals but are not related to either the real or complex numbers.

In the early 1970s John Horton Conway, a British mathematician who has made major contributions to many areas of mathematics and is best known for his invention of the cellular automaton game called *Life*, constructed a continuum of numbers now called *surreal numbers* that not only contain the real numbers but also contain numbers *larger* than any real number as well as numbers *smaller* than any real number.

Andrea Bruder is Assistant Professor of Mathematics at Colorado College.
Andrea Easterday received her Ph.D degree in 2011 from the University of California at Santa Cruz.
John J. Watkins is Professor Emeritus of Mathematics at Colorado College.

Bibliography

[1] A. Benjamin and E. Brown (eds.), *Bisquits of Number Theory*, The Mathematical Association of America, 2009.

[2] A. Beutelspacher, *Albrecht Beutelspachers Kleines Mathematikum. Die 101 wichtigsten Fragen und Antworten zur Mathematik*, C. H. Beck, 2010.

[3] A. Beutelspacher, *Cryptology*, Spectrum, 1994.

[4] D. Burton, *Elementary Number Theory*, 7th edition, McGraw-Hill, 2010.

[5] H.-D. Ebbinghaus et al., *Numbers*, Springer, corrected edition, 1996.

[6] R. Friedberg, *An Adventurer's Guide to Number Theory*, Dover, 1995.

[7] R. Hill, *A First Course in Coding Theory*, Clarendon Press, 1990.

[8] G. Ifrah, *The Universal History of Numbers*, Wiley, 2000.

[9] K. Menninger, *Number Words and Number Symbols*, Dover, reprint edition, 2011.

[10] C. Reid, *From Zero to Infinity: What Makes Numbers Interesting*, 5th edition, A. K. Peters, 2006.

[11] P. Ribenboim, *My Numbers, My Friends*, Springer, 2000.

[12] J. Watkins, *Number Theory: A Historical Approach*, Princeton University Press, 2014.

[13] D. Wells, *The Penguin Dictionary of Curious and Interesting Numbers*, Penguin, 1986.

[14] H. Wußing, *6000 Jahre Mathematik. Von den Anfängen bis Leibniz und Newton*, Springer, 2008.

[15] H. Wußing, *6000 Jahre Mathematik. Von Euler bis zur Gegenwart*, Springer, 2009.

Index